Nuclear Energy Today and Tomorrow

A course of lectures on selected topics in the fields of nuclear and atomic energy

D. Z. ROBINSON
Vice-President of Academic Affairs,
New York University, New York

C. B. A. McCUSKER
Professor of High Energy Nuclear Physics
University of Sydney

P. W. McDANIEL
Director of Research,
U.S. Atomic Energy Commission,
Washington, D.C.

W. K. H. PANOFSKY
Director,
Stanford Linear Accelerator Centre,
Stanford University, California

R. H. DALITZ
Royal Society Research Professor,
Oxford University, England

Heinemann Educational Books
London and Edinburgh

Heinemann Educational Books Ltd

LONDON EDINBURGH MELBOURNE TORONTO
SINGAPORE JOHANNESBURG AUCKLAND IBADAN
HONG KONG NAIROBI NEW DELHI

IN THE SAME SERIES

Pioneering in Outer Space

ISBN   0 435 68280 6

© Shakespeare Head Press, Scientific Services Pty. Ltd. and
S. T. Butler
First published in Great Britain 1971

O 8166

ↄ

Published by Heinemann Educational Books Ltd
48 Charles Street, London W1 X 8AH

Printed in Great Britain by
Biddles Ltd., Guildford, Surrey

PREFACE

The Science Foundation for Physics within the University of Sydney is honoured to present such a distinguished group of lecturers at its 12th International Science School for High School Students.

On behalf of the Foundation we wish to take this opportunity of thanking Professor R. H. Dalitz, Professor C. B. A. McCusker, Dr. P. W. McDaniel, Professor W. K. H. Panofsky and Dr. D. Z. Robinson, for having given so generously of their time and effort.

We have chosen the general heading, "Nuclear Energy Today and Tomorrow", for the 1969 International Science School because its lecture course covers selected topics in the fields of nuclear and atomic energy. In each of the fields discussed the lecturers are specialists of world renown and the material has been specially prepared, written and edited for fifth-year high school students. We therefore feel that the lectures will be of interest not only to the students, but to the widest sections of the public. We feel that the material presented will be generally appreciated by the increasingly more science-conscious layman in this scientific age and also, in fields other than his own, by the specialised scientist.

The Foundation's 1969 International Science School and, indeed, this book, are intended to stimulate and develop science consciousness, in Australia and throughout the world. The Foundation is therefore honoured that, once again, under a special scheme, 20 students from the United States, Britain and Japan—specially selected in their respective countries for their ability and scholastic success—are attending the Science School in Sydney. These students have been designated respectively "The President's Australian Science Scholars", the "Royal Institution Australian Science Scholars" and the "Sato Eisako Australian Science Scholars".

The 20 overseas students, as well as those selected from throughout Australia and New Zealand deserve the applause of all, and the Science Foundation for Physics within the University of Sydney is happy to honour and reward the ability and diligence of these young people.

Sydney, August, 1969   H. MESSEL and S. T. BUTLER

INTERNATIONAL SCIENCE SCHOOL SERIES

Edited by

H. MESSEL

B.A., B.Sc., Ph.D.
Professor of Physics and
Head of the School of Physics
University of Sydney

S. T. BUTLER

M.Sc., Ph.D., D.Sc.
Professor of Theoretical Physics,
University of Sydney

THE SPONSORS

The Science Foundation for Physics within the University of Sydney gratefully acknowledges the generous financial assistance given by the following group of individual philanthropists and companies, without whose help the 1969 International Science School for High School Students and the production of this book would not have been possible.

FULL SPONSORS

Ampol Petroleum Limited

The James N. Kirby Foundation

Pergamon Press Limited

The Sydney County Council

PART SPONSORS

A. E. Armstrong, Esq.

A. Boden, Esq.

The Commercial Bank of Australia Limited

Conzinc Riotinto of Australia Limited

The Nell and Hermon Slade Trust

Standard Telephones & Cables Pty. Ltd.

CONTENTS

Science and Society
D. Z. ROBINSON

The Peaceful Use of the Atom
P. W. McDANIEL

Cosmic Radiation
C. B. A. McCUSKER

Particle Physics
W. K. H. PANOFSKY and R. H. DALITZ

Science and Society

(Two Chapters)

by

DAVID Z. ROBINSON

Dr. David Z. Robinson

Vice-President of Academic Affairs,
New York University, New York.

CHAPTER 1

Science and Society

The purpose of these lectures is to describe some of the present-day interaction of science and government, both with regard to the use of science to further governmental purposes and the use of government to support scientific activities.

My topic today will be the past and present role of science in government. My next lecture will concern the role of government in science.

Ever since the war machines of Archimedes were enlisted in the defence of Syracuse, the work of scientists has been put to the use of their governments. In the old days, of course, governments were primarily concerned with war and currency. Newton spent a good part of his life as director of the Mint in Britain. (His job was not a sinecure but one involving a great deal of his time.) Lavoisier worked on the Gunpowder Commission for the French Government before the Revolution. Charles II set up the Royal Observatory of Greenwich in 1714 to seek better navigational aids for the Royal Navy and the merchants of London. That same year Parliament offered £20,000 for invention of a device for determining longitude within half of a degree.

A. Hunter Dupree,[1] the best historian of relations between science and government in the United States, has described the impact of the American Civil War of 1861-1864 on the government's ability to use science:

---

1 A. Hunter Dupree, *Science and the Federal Government,* Harvard University Press (1957), page 120.

"In the 'last of the old wars and the first of the new', the battles on land were not impossibly different from those of the Napoleonic era, which served as models for the officers of both sides.   And yet, the introduction during the years of peace of the railroad, the telegraph, and innovations in the construction of guns had changed the pace of warfare and upset the balance of offence and defence, producing a new pattern in which the old rules, based on a relatively static technology, offered poor guidance.   At sea, where technology had affected the methods of warfare more deeply, the change was dramatic."

The large-scale construction of ironclad ships with revolving turrets; the enlistment of Joseph Henry, Director of the Smithsonian Institution and America's best physicist, in the development of observation balloons; advances in medical organisation, such as field hospitals and mass medical examinations — marked the beginning of the varied modern links between science and warfare.

As governments have expanded their activities, their involvement with science and technology has also expanded.   In the United States, the Geological Survey and the Coast Survey were the first scientific organisations.   The United States Government set up the Department of Agriculture in 1862.   The Department worked on plant and animal diseases, hired scientists, set up an organisation to disseminate research results to farmers.   In its understanding of the *total* need of the farmer for research, and its growth from doing pure research to adding applied research, development, education and demonstration, the Department practised what is now called the "systems approach".

The growth of industry in the United States also led that government at the beginning of the 20th century to develop agencies which would be helpful to industry.   Thus although the United States Constitution calls for government responsibility for maintaining standards of weights and measures, Congress only established a separate organisation for this purpose in 1901 (following the pattern in Germany of the Physikalische-Technische Reichsanstalt and in England of the National Physical Laboratory).   A permanent Bureau of the Census, a Bureau of Mines, and the National Advisory Committee for Aeronautics showed government interest in civilian industry as well.

Similarly in the past 50 to 100 years governments have been

increasingly involved in health care. They have supported research in improving health and attacking diseases as a regular activity although until recently the charge could be made that more money was spent by the Federal Government on eliminating diseases in animals than in humans.

As the role of science in government has grown, so has the role of the scientist. Scientists act as government *employees* working on specific technical projects; they act as *advisers* to policymakers, and—more and more in recent years—they act as *policymakers* directly responsible for government programmes. The structure and use of science and scientists in these three roles has become increasingly formalised in the United States. There are signs that other countries are trying to change their structures to more closely approach the American model.

The scientist as an employee can work in any one of a number of government laboratories. Thus scientists are found all through the space agency and are heavily involved in developing and supervising the rockets and systems that make up the space programme. Scientists man the Agriculture Department laboratories aimed at improving the quality and quantity of food. They develop fish protein concentrate for the Bureau of Fisheries, programme the computers predicting weather patterns, test new aircraft designs in wind tunnels, develop theories of combustion in the Bureau of Mines, measure time to an accuracy 1 part in 10^9 in the Bureau of Standards, and study spectroscopy of paint samples in the F.B.I. Scientists work directly in medical research in National Institutes of Health laboratories, and also supervise the agency's support of health research outside the government.

The civil service scientists in the Federal Government are a very powerful group. They have expertise which the Congressman knows he does not have, and this gives them independent strength in dealing both with Congress and with their own "civilian" superiors. Congressmen believe that they understand social science or management and are thus willing to disagree vigorously with professionals in this area. However, when the Director of the Geological Survey, the Bureau of Mines, or Bureau of Commercial Fisheries comes before a committee of the Congress to justify his programme, he will usually be treated very gently—more gently than his political superior, the Secretary of Interior.

The scientist as adviser is not altogether new, but it is a role that has grown greatly in importance. The National Academy of Sciences was established during the Civil War primarily to give scientific advice to the U.S. Government. During its first 50 years of existence—until 1913—the Academy was asked by the War Department for advice on exactly five matters: "On the Question of Tests for the Purity of Whiskey"; "On the Preservation of Paint on Army Knapsacks"; "On Galvanic Action from Association of Zinc and Iron"; "On the Exploration of the Yellowstone"; and "On Questions of Meteorological Science and its Applications."[2] During World War II, however, scientists were brought in at all levels to give advice to the government and the success of this advice in the view of the policymakers has led to the appointment of full-time advisers. There are now science advisers to the Secretary of State, the Secretary of Interior and the Secretary of Agriculture, as well as to the President of the United States.

The science adviser or advisory committee has four main functions in a government context, regardless of the particular subject area:[3]

(1) He analyses the technical aspects of major policy issues, and interprets them for policymakers.

(2) He evaluates specific scientific or technological programmes and provides advice on organisation and budget.

(3) He identifies new opportunities for research and development.

(4) He helps in the selection of individual research proposals for support. (This kind of advice is usually given to the project officers of a government agency that supports science.)

More recently scientists have been given direct responsibility for making policy in some government agencies. There is an Assistant Secretary of Commerce for Science and Technology, who supervises the Patent Office, the National Bureau of Standards, and the Environmental Science and Services Administration. The Chairman of the Atomic Energy Commission, Dr. Glenn Seaborg, the Director of the National Science Foundation, Dr. Leland

2 Warner Schilling in *Scientists and National Policymaking,* edited by Robert Gilpin and Christopher Wright, Columbia University Press (1964), page 145.
3 For a detailed discussion see Harvey Brooks, "The Scientific Adviser" in Gilpin and Wright, ed., *Op. Cit.* page 73.

Haworth, and the Deputy Director of Research and Engineering
of the Department of Defence, Dr. John Foster, are all scientists.
The Secretary of the Air Force from 1966-68, Dr. Harold Brown,
was a physicist.

In the United States, since 1957, the chief adviser on science
in government has been the Special Assistant to the President for
Science and Technology, popularly known as the President's
"Science Adviser".   President Eisenhower chose James R. Killian,
then president of Massachusetts Institute of Technology, and later
George B. Kistiakowsky, a Professor of Chemistry at Harvard, as
advisers.   Jerome B. Wiesner of M.I.T. was President Kennedy's
Science Adviser, and Donald F. Hornig, a Professor of Chemistry
at Princeton, was President Johnson's.   President Nixon has chosen
Lee DuBridge, president of the California Institute of Technology,
as his adviser.

The science adviser plays a key role in all aspects of science
and public policy and the mechanisms that he uses are varied.
Basically he has four jobs:

1. Chairman of the President's Science Advisory Committee (PSAC).

This group, made up of non-government scientists chosen from
all over the country, meets formally two days a month and
reviews a wide range of scientific problems.   These might include
future of the space programme, the status of the nuclear non-
proliferation treaty, the world food problem, weather modification,
or the health of basic research in biology.   When the President
or PSAC itself sees a problem as particularly important and timely,
the committee sets up a special panel of experts to consider the
problem in depth.   These panels report back to the President,
and sometimes their report is published.   Panel Reports have
been published on such subjects as the World Food Problem; the
Use of Computers in Education; the Future of the Space Programme; Science, Universities and the Federal Government; and
Environmental Pollution.

Although PSAC serves at the pleasure of the President, there
have been no major changes in personnel (other than the normal
rotation of members) when new Presidents have come into power.

The committee looks on its primary job as adviser to the President, but the reports it turns out can have an influence on agency programmes or on public attitudes on various subjects. Although the membership may gradually change its character according to the interest of the President in specific problems (Eisenhower was primarily interested in using the committee in areas of national security; Kennedy was interested in civilian aspects of technology as well and Johnson had a great interest in water and medical sciences), the committee has remained a remarkably non-partisan group of people whose advice has been seriously considered by the four Presidents for whom they have worked.

2. Chairman of the Federal Council for Science and Technology (FCST)

The Federal Council, founded in 1960, is made up of the top scientific officials in the Federal Government. FCST's job is to co-ordinate scientific programmes and policies *inside* the Federal Government, and it operates primarily through special committees. For example, there is an Inter-agency Committee on Atmospheric Sciences (ICAS) which reports to the Federal Council. This committee includes members from all agencies interested in the field of meteorology or atmospheric science: the National Science Foundation (which supports atmospheric sciences), the Weather Bureau (which predicts the weather), the Federal Aviation Administration (which controls aircraft), the military services (whose men operate in and under the weather), the Atomic Energy Commission which is concerned about fallout), the Department of Agriculture (which wants rain), the Department of Interior (which is interested in weather modification as it affects water supply) and the Space Agency (which studies the upper atmosphere). Each of these agencies has a specific research programme. The committee's job is to review the programmes, to search for possible duplication, to find gaps that need additional research, and to exchange information. It writes an annual report reviewing and summarising the government effort in atmospheric sciences. These reports are the only place where such information can be found. As a practical matter, the Federal Council cannot resolve differences between agencies, but it can lay these differences out in the open so that they can be adjudicated or ignored.

3. Director of the Office of Science and Technology (OST).

This office was established in 1962, when the staff of the President's adviser was moved from the President's personal staff into the Executive Office of the President.

One major reason for this change was to allow the President's Science Adviser—in his new role as Director of OST—to testify before Congress on matters involving science co-ordination and science programmes. The tradition of separation of powers in the United States has meant that presidential advisers do not normally testify before the Congress without special permission from the President. The Science Adviser still does not testify on matters in which he advises the President directly. The staff of OST forms the secretariat of PSAC panels, sits on FCST committees, and advises the Bureau of the Budget on technical aspects of government programmes. Since a budget for a scientific programme determines what will be carried out, the advice to the Budget Bureau can have a major impact.

4. Special Assistant to the President for Science and Technology (SAPST).

In his role as Special Assistant, the Science Adviser deals with issues in which the President is personally involved, including foreign policy issues. President Kennedy sent Jerome Wiesner to Pakistan to study the problem of "water-logging". Over the years, water from irrigation ditches had raised the watertable so high that the salt the water brought along with it stopped the grain from growing. Wiesner visited Pakistan and later sent a group of scientists to recommend solutions. When protests were made by the Argentine Government over the banning of Argentine beef because the U.S. Department of Agriculture had found that the virus which caused hoof-and-mouth disease could theoretically survive standard curing methods, Kennedy asked Wiesner to look into it. Wiesner set up a special PSAC Panel under Dr. George Harrar of the Rockefeller Foundation. This Panel studied the problem and recommended procedures that would let cured beef be imported safely. It also recommended a research programme that would help Argentina find longer range solutions. Thus a potentially tricky problem in our foreign relations was alleviated.

When President Johnson was looking for a way to improve U.S.-Korean relations (at the time of President Park's visit in 1964), he turned to Donald Hornig for suggestions. As a result, Johnson was able to announce that he was sending a team led by his personal science adviser, to see how industrial research in Korea could be improved. From that trip came a programme for a Korean Institute of Science and Technology that was successful in attracting Korean scientists back from the United States to Korea.

During this same time, two emotional issues with scientific overtones were complicating our relations with Europe. The first was the "brain-drain": the immigration to the United States of scientists from other countries. (This issue caused a debate on the floor of the British Parliament.)

The second was the "technological gap" between the United States and Europe. According to the believers in the gap, the United States is now far ahead of Europe in some modern fields of technology—such as electronics, computers and aircraft—and this is leading to an unhealthy relationship in which Europe will become dependent on the United States. Dr. Hornig took the lead in fostering studies of these issues and bringing them up for discussion in international meetings, particularly those of the Organisation for Economic Co-operation and Development (OECD). The net result was greater understanding by both sides and a significant cooling off of the issues as a political force.

The Special Assistant can also be called in on domestic issues by the President. For example, when the north-eastern part of the United States was blacked out due to a power failure in 1966, the President immediately called Dr. Hornig and asked him to look into the matter. Hornig got a committee together and this committee made recommendations on national supervision of power networks.

The Nuclear Test Ban Treaty:

A Case Study of Science in Government

In the late 1950's, American, British and Soviet scientists were involved in discussions at Geneva that led eventually to the nuclear test ban treaty. These negotiations indirectly made scientists into diplomats, a role they had difficulties in fulfilling well. The story shows some of the limitations as well as some of the strengths of scientists in political roles.

The atomic bomb was developed in 1945, and in 1950 the U.S. made a major decision to go ahead with the development of the hydrogen bomb. Scientists were deeply involved in this decision and it took a heavy toll of their unity and political innocence. The acrimony of the then-secret discussions as to whether or not the H-bomb should be developed, began afresh in 1954 with the trial of Robert Oppenheimer. In retrospect it seems clear that the intensity and divisiveness of the debate among scientists over nuclear fallout were related to the personal animosities engendered by the trial itself.

In any event both the Soviet Union and the United States developed the hydrogen bomb in the early 1950's. The first thermonuclear device was exploded by the United States on Eniwetok Atoll, November 1, 1952. In March, 1954, seven months after the first Russian hydrogen explosion, the United States exploded its first "dropable" hydrogen bomb which had the energy of 15 million tons of TNT. Part of the debris from this explosion went into the upper atmosphere and fell on a Japanese fishing ship, *The Lucky Dragon*, which was about 100 miles away. The fishermen became sick from the radioactivity and contaminated radioactive fish from the test area began to reach the Japanese market. The problem of short-range and long-range "fallout" was raised very severely for the first time as a political issue.

No one had known very much about fallout before this, but American biologists had been studying its effects. Before the explosion, it had been expected that there would be some short-range fallout right around the nuclear explosion. It had been assumed that most of the radioactive debris that did not drop immediately would be sent into the upper atmosphere and carried around the world. It would gradually fall down and, in general, it would have very little effect of a biological nature.

The major reason for lack of concern with fallout was that we are already surrounded by a background of radioactivity both from natural radioactivity in rocks and from cosmic rays. Calculations seemed to show that the fallout from nuclear testing was a small fraction of this natural background (except in extraordinary circumstances like *The Lucky Dragon*). Furthermore, it was thought that the effects of radiation were negligible unless a certain "threshold" of radioactivity was reached. However, in the late 50s, new evidence from laboratories seemed to show that genetic mutations increased proportionately to the increase in radiation, and there was *no* threshold. They also seemed to show that in addition to genetic effects, which, of course, affect only unborn generations, there were somatic effects—such as increase in leukemia rate—which might also be proportional to the amount of radiation. Thus, from a technical point of view, it appeared that even though the natural background was increased only slightly from the fallout due to nuclear testing, still the effects would be bad. In addition, it appeared that certain fission products (such as strontium-90) were preferentially absorbed by children. In any case, no one could claim that fallout was good for you, although some were accused of trying to make that attempt.

A world-wide effort was started in the middle and late '50s to oppose nuclear testing in order to eliminate radioactive fallout. This led to a bitter public debate between scientists—one of the first of its kind. The chief figures in this debate were Linus Pauling, Professor of Chemistry at California Institute of Technology, and Professor Edward Teller from the University of California at Berkeley. These men came to completely different conclusions from the same scientific evidence and neither of them can be accused of falsifying the evidence. Gilpin[4] gives an example of the way each of the men used the same data on the harmful effects of fallout. Pauling would present his estimate of the total number of individuals who would die from a given amount of fallout. This number would not be compared to the total number of deaths that would occur due to the natural radiation background—a figure 50 to 100 times as great. Teller, on the other hand, would present the data *only* in comparative terms. That is to say, he would average the number of individual deaths due to fallout in terms of days of life lost by the American people as a whole. He would then compare these to the days lost due to smoking. Teller would also point out that people who live in

---

4 Robert Gilpin, *American Scientists and Nuclear Weapons Policy,* Princeton University Press (1962), page 167.

Denver receive more background radiation than people who live in Los Angeles, because of the greater number of cosmic rays at high altitudes. The increase in radiation in Los Angeles due to fallout from any projected amount of testing was no greater than the increase that one would get from moving from Los Angeles to Denver.

The debate over nuclear testing and fallout was not confined to the scientific community. In 1954, Prime Minister Nehru of India made what was probably the first proposal that nuclear testing be stopped as a step toward disarmament. During the 1956 election campaign, Adlai Stevenson favoured a test ban and President Eisenhower opposed it. The next two years saw a gradual change of administration policy. The public outcry over fallout, the rising cost of atomic testing, and President Eisenhower's desire to move towards disarmament were all factors in this change.

In 1957, James Killian was appointed as the first full-time Science Adviser to the President. The President's Science Advisory Committee assumed its present form and White House status. Possessing the important advantages of objectivity, technical knowledgeability, lack of identification with either the strong proponents or opponents of nuclear testing, and access to the President, the committee played a strong role in the government discussions between the various agencies with regard to disarmament.

In January, 1958, President Eisenhower wrote a letter to Premier Bulganin of the Soviet Union proposing technical studies of the inspection and control measures necessary for possible future disarmament. He presented a "package" of several items for possible study, including a nuclear test ban.* There was no answer to the letter.

At about this same time, the President instructed Killian to appoint a special panel of experts to study the effects on national security of a nuclear test ban and the technical feasibility of a control system to monitor such a ban. Killian made Professor Hans Bethe of Cornell the chairman of the panel. The Bethe Panel, which included representatives from the Defence Department and AEC as well as non-government scientists, reported in April, 1958, that a test ban would not prejudice national security. Their report

---

* Other items were prevention of surprise attack, cessation of production of nuclear materials and banning of weapons in outer space.

—together with the announced Russian suspension of testing in March—influenced the letter that the President sent to the new Premier, Khrushchev, in April, renewing his proposals for technical discussions on disarmament inspection systems. The letter seemed to indicate for the first time that the United States would agree to a nuclear test ban separate from a general arms control "package", provided that the Soviet Union would agree to a workable inspection system. The Russians responded favourably to the idea of discussions on the test ban alone.

On July 1, 1958, a group of technical experts convened in Geneva. The conference was officially known as the Conference of Experts to Study the Possibility of Detecting Violations of a Possible Agreement on Suspension of Nuclear Tests. The meeting appeared to be purely technical—a gathering of British and American scientists with their technical counterparts from the Soviet Union. Indeed, there were many scientific and technical issues which had to be discussed. In truth, however, the technical issues were intimately intertwined with political ones, and the scientific discussions turned into negotiating sessions for which the American scientists were not fully prepared.

Gilpin gives various examples of Russian manoeuvring for political advantage in the "technical" talks. One was an apparently innocent suggestion by the head of the Russian delegation that the language which permitted routine control commission flights to collect radioactive debris "over the open seas" be amended to add the clarifying parenthesis "(oceans)". "Legally", Dr. Federov explained, "everything may be considered to be an open sea—but, practically speaking, such flights will take place over oceans". Dr. Fisk, leader of the U.S. delegation, amiably suggested that it would save words to simply say "over the oceans" and the Russians hurried to endorse and sew up the agreement. While Fisk sought no legal counsel on the implications of this change, it could have made control flights near the Soviet Union impossible because the U.S.S.R. is bounded by several "seas" and no "oceans" except at the lower extremity of the Kamchatka Peninsula.[5]

The main technical issues before the experts' conference were two: first, could nuclear explosions be *detected*, and second, could they be *identified*, that is could they be distinguished from natural

5 Gilpin, *Op. Cit.*, pages 211-212.

events such as earthquakes that also occur. For the U.S. a major issue was whether and how there could be verification *through inspection* of suspected violations of a test ban agreement. For the Soviet Union, which traditionally has relied very heavily on secrecy, the major issue was whether all detection and identification of possible tests inside Russia could not be done from *outside* their borders.

It turns out that the detectability and the identifiability of nuclear explosions vary tremendously depending both on the size of the explosion and the medium in which it takes place. The conference agreed, for example, that explosions as small as one kiloton in the lower atmosphere could be detected over long distances by electromagnetic pulses and could be identified by sampling the radioactivity in the atmosphere for nuclear fallout. They also agreed that detection of test shots in the ocean would be relatively easy because of the size of the disturbance they would cause, but that positive identification of underwater explosions would be much more difficult.

Nuclear explosions underground above a certain size could be detected with sensitive seismographs over long distances. (In fact, one of the few pieces of data available to the scientists at the conference was that a $1·7$ kiloton atomic explosion had been detected 2,300 miles away.) However, it would still be difficult to distinguish between a nuclear explosion and an earthquake in the middle of Siberia or Colorado from outside the borders of the country. The best system of identifying an underground nuclear explosion would be to go to the suspected site and drill to see if radioactive material is present.

Nuclear explosions in outer space, the scientists agreed, would be very difficult to either detect or identify although they felt it would be possible to use some satellite techniques.

The main technical problem that the Geneva conferees discussed was the development of a system that could detect underground nuclear explosions. As George Kistiakowsky later noted, the conferees found "a dearth of experimental data on which to base their conclusions". They found that seismological research was very primitive and very little was known about the development and propagation of seismological signals from explosions. There are theoretical ways by which an explosion could be distinguished from an earthquake. For example, the earthquake is usually caused by slippage and that means that on one side of the quake its first movement is outward of the earth and on the other side, it is inward, whereas from an explosion, the motion is outward in all directions. Sensitive seismographs might detect such a difference if they were close enough to the event. But it was not clear how close the seismographs would have to be.

Nevertheless, the scientists felt a commitment to develop or design some detection system. By the end of August, the Geneva scientists came forth with a system design which they felt could detect one kiloton explosions in the atmosphere and detect 5 kiloton explosions underground. The system involved setting up about 180 control posts throughout the world which would make seismological recordings and monitor radioactivity.

The report was received with world-wide acclaim. Here was the model for resolution of East-West differences: technical negotiations between scientists from both sides!

The heady atmosphere of success did not last long. There were technical qualifications attached to the original report regarding the system's limitations and the lack of essential data. In their enthusiasm for the report, opponents of testing had ignored these qualifications—but they were soon in the spotlight. Dr. Teller and a number of other scientists emphasised different ways of reducing the probability of detection. Further seismological studies gave greater understanding of the technical difficulties. When political negotiations started in October, there were many arguments with the Soviets as to what the appropriate number of inspections had to be for a system to be satisfactory. (The United States announced a moratorium on tests at that time.) The political negotiations dragged on for more than two years. In 1961 the Soviet Union broke their self-imposed moratorium on testing, and the United States started a series of tests somewhat later. Finally, in 1963—five years after the scientists met in Geneva—agreement was reached on a test ban in the atmosphere only. Underground testing was allowed to continue, and the system of control posts was never put into effect.

It is easy to overstate the "failure" of the Geneva Conference of experts. The agreement which came forth represented important concessions on both sides. In particular, the Soviets agreed that some kind of inspection was needed. The conference began the modernisation of research in the whole field of seismology and led—in time—to considerable improvement in detection capability.

Nevertheless, the detection system was never adopted, and in fact the emphasis on technical issues may actually have handicapped the process of political agreement. Skolnikoff[6] has written:

> . . . How can U.S. demands for assurance of compliance with the treaty be squared with Soviet needs to maintain secrecy? Are there ways of devising means for operating a control commission that would be accepted as neutral by both sides? Would disputes arising under the treaty undermine the very increase in trust that was considered one of the purposes of the agreement? Of the greatest significance of all, is there enough of a genuine community of interest between Russia and the United States to make an agreement requiring international inspection and control viable?

> These questions did receive attention, but the over-emphasis on the technical solution resulted in consternation and confusion whenever clever new technical ideas were invented whereby a nation conceivably might be able to evade the control system. Some of these new technical ideas were important, but they rarely were given their proper weight in relation to the basic political issues of the negotiations. Instead, they caused the attention and effort of the government to be diverted to a consideration of the marginal technical gleam in someone's eye rather than to remain focused on the central issues."

The atmospheric test ban of 1963 was a very good first step but it was *not* disarmament. However, since the fight against nuclear testing had been based on the danger of fallout, public pressures for disarmament or even the complete stoppage of tests were greatly reduced once the danger of fallout was essentially eliminated.

The final irony of the long divisive debate over nuclear testing was that it concerned the wrong issue. The real reason for having a test ban or indeed disarmament arises from the danger of nuclear war. No matter what the estimate of damage from fallout, this damage is trivial compared to that which would result from nuclear

---

6 Eugene B. Skolnikoff, *Science, Technology and American Foreign Policy*, M.I.T. Press (1967), page 114.

war.   In fact, people on both sides of the test ban issue knew this.   The arguments for stopping testing because of the fallout were made by people who really opposed testing because they thought that it would increase the probability of nuclear war.   The argument which belittled the effects of fallout was made by people who felt that only through testing could we assure the strongest possible arsenal and thus deter a war.   Yet, neither side really put the argument in that form, and the chance to engage public discussion on this key issue of preventing war—rather than over fallout and testing—was lost.

Conclusion

Despite the political dangers scientists were essential participants in the discussions that led to the test ban.   The solution to the problems the U.S. had in Geneva is to have a clearer understanding of both the technical and political factors by both scientists and politicians.

In the future we will need more—not less—science and technology in government.   If wise national policy is to emerge, it will require the services of first-class scientists.   I would like to see all young scientists consider several years of government service to be part of their education and their social obligation.

CHAPTER 2

Society and Science

We have talked some about the long relationship between science and government in terms of the use of science and scientists in the work of the government itself. Another part of this relationship is the interest that government has taken in the work and health of science.

The U.S. Constitution empowered Congress to "promote the Progress of Science and the Useful Arts, by securing for limited times to Authors and Inventors the exclusive Right to their respective Writings and Discoveries". As this phraseology indicates, government interest in furthering invention and technology began very early. It was a long time, however, before governments discovered that *scientific* advance — progress in basic research — was also vital to the national well-being.

Right up until the Second World War, government shared in the general public consensus that while industry could use engineers and applied scientists effectively, society could afford to support pure scientists and their work only insofar as they were educating students at the universities.

As an example of this type of thinking, at the beginning of World War I, Thomas Edison was appointed chairman of the Naval Consulting Board charged with finding and promoting new inventions to further the war effort. Edison told the Navy that they should hire at least one physicist just in case they had to "calculate something".[1]

World War II came 25 years later. The American and British Governments were desperately trying to develop radar for the

---

1 R. Gilpin, *American Scientists and Nuclear Weapons Policy,* Princeton (1962), page 10.

location of enemy aircraft. Radio engineers in Britain and the United States had done a good deal of work on the problem. But in the end, the key breakthroughs in the development of short wavelength devices came not from the engineers who had had experience with long wavelength devices but from physicists who understood the fundamentals of electromagnetic microwave theory, and could understand microwave propagation.

Daniel Greenberg interviewed some of the top people who worked in the famous MIT Radiation Laboratory that produced military radar.[2] I. I. Rabi, a physicist who served as associate director, observed, "The people at the Rad Lab didn't know a damn thing about radar. I was in charge of a magnetron group and I'd never seen one. Well, I thought, I'll go around to MIT and ask some of the electrical engineers. After talking to them, I could see they didn't know anything either . . . so we started absolutely fresh and designed magnetrons". "(The physicists) were light on their feet," was the explanation of Luis Alvarez, a physicist who served both at the Rad Lab and in the Manhattan Project. "They knew something about electronics because of their work with accelerators, but the real reason was that they were the best people and they were adaptable to anything."

Such assessments of the wartime experience may partake of scientific chauvinism. But the physicists *were* "light on their feet" and they succeeded in what seemed an impossible task.

Radar and the proximity fuse were tremendously important, but the major military development of World War II was the nuclear bomb.

Throughout the 1930s, physicists had been working with radio activity and trying to understand quantum theory — on university time pretty much, and without any government support. No one — not even the scientists— expected that the work would ever have practical applications. The famed Cambridge professor, Ernest Rutherford, said in 1933, "Anyone who expects a source of power from the transformation of these atoms is talking moonshine".

In 1939, nuclear fission was discovered. Within five years, this basic scientific discovery changed the course of world history.

---

2 Daniel S. Greenberg, *The Politics of Pure Science,* New American Library (1967), pages 90-91.

It revolutionised the military and political positions of the world powers and it offered an abundant new energy source to supplement the supply of fossil fuels.

Three aspects of the bomb development experience served to convince the U.S. Government — more or less permanently — that basic research was at least as important as applied research and deserved continuing government support:

- First, the basic research led very quickly to the development of a major military resource.

- Second, the standard approach of applied research — in this case, setting out to develop a powerful explosive — would never have produced a nuclear bomb since the people who were experts in explosives didn't know any nuclear physics. Undirected research, performed because of the internal logic of the science itself, proved to be more important than directed research.

- Third, a storehouse of basis knowledge was the real prerequisite for the bomb's development.

Scientists learned something, too, from the Manhattan Project and Los Alamos experience. This large development programme depended on a base of scientific knowledge, but the development activities also *contributed* to that base. These projects produced a great deal of valuable scientific information.

The discovery of nuclear fission produced more than a new source of explosives. Scientists had found a temporary solution to the problem of finding sufficient energy to run the machines of the world. (*See Table I.*)

This discovery came at a time when many people thought we would soon be running out of oil and other fossil fuels.

Physics was not the only field in which science was producing "miracles". The discovery of penicillin had come from basic research in the characteristics of moulds, and sulfa drugs were the product of chemical research. New antibiotics, such as auriomycin and terramycin, were being discovered at a rapid rate.

Suddenly basic research — the remote and frivolous stepchild of "Science and the Useful Arts"— became the favourite son. It had alleviated some of the country's deep fears: the fear of military defeat, the fear of inadequate energy resources, and the fear of disease. College and Community Chest fund-raisers — as well as

TABLE I.—Comparison of U.S. energy resources with projected requirements, 1960-2000.
(Expressed in terms of energy content, where the unit is 1 Q = 1 quintillion (10^{18}) Btu)

	Coal	Conventional liquid hydrocarbons	Shale oil	Oil in bituminous rocks	Natural gas	Uranium	Thorium
1. Known reserves recoverable under present economic and technologic conditions	4.6	0.3	....	....	0.3	0.3	....
2. Undiscovered deposits recoverable (when found) under present technologic conditions	....	1.3	....	....	1.3	0.8	....
3. Additional known and undiscovered resources possibly recoverable in the future	84	2.3	945	0.07	0.9–25	224,000	336,000
4. Total resources (sum of lines 1, 2 and 3)	88.6	3.9	945	0.07	2.4–26	224,000	336,000
5. Cumulative projection of energy requirements 1960-2000, by source	0.6	1.4			1.0	0.2	

Source: *Energy R&D And National Progress*: U.S. Government Printing Office, 1956.

Congressmen — know that people give money more readily to get rid of fears than they do for more positive purposes. Such was the case in supporting science research.

The scientists who left the military laboratories at the end of the war to return to universities found that they would need costly new tools to advance their research. The Federal Government — particularly those parts dealing with military and health problems — was willing for the first time to give direct support to basic research. The question was how to get together. There had been an exciting wartime romance between science and government. But would it make a happy marriage? The post-war period of government scientific support was dominated by practical and procedural questions: How much money should be spent on science? Which federal agencies should spend it? How should research support be controlled? What form should it take? Who decides whether a project is worthwhile?

For about fifteen years, these questions were answered in a rather easy-going and trusting atmosphere. More recently, the role of government in science has departed rather sharply from the honeymoon phase. Important and difficult questions of resource allocation are being raised and in a kind of hard-knocks political arena. I will return to these questions.

In 1945, the problem seemed relatively straightforward. The Federal Government was convinced that it should support basic science and science education in a "civilian" context — that is, primarily in universities — but it had no mechanism for doing this.[3]

While there was considerable discussion about setting up a special agency to support science, it was clear that this would take a long time. Meanwhile, three agencies separately stepped in, each with its own goals and its own — essentially practical — interest in scientific research.

The Atomic Energy Commission, the successor to the military agency that developed the atomic bomb, began supporting nuclear

3 Historically, in the United States, education at the university level has been the responsibility of states or private institutions. The Federal Government was involved in just two ways: The Government was supporting its military academies and a few specialised institutions. It also was giving some support on a matching basis to the so-called land-grant institutions in each state for agricultural research and the dissemination of information on modern agricultural techniques.

research at universities and in industrial laboratories. The Commission was forbidden by law to have laboratories of its own, so it had to learn to deal with outside institutions in order to do the research that it wanted accomplished.

The Defence Department had been dealing very successfully during the war with outside laboratories attached to universities.[4] The Navy saw the need of continuing research support and developed a flexible contract form for contracting with universities for basic research in oceanography and other fields related to its military mission. There was some opposition at the time to military research in the universities. Scientists were tired of military control. However, the Navy showed that it was only interested in supporting good science, and most scientists accepted basic research contracts from the Navy.

Finally, the National Institute of Health — which was set up primarily to do health research in its own laboratories — started to give grants for research in areas related to the cure of specific diseases. Most of the grants were to medical schools.

While grateful for these multiple Federal bounties the scientific community remained concerned about their source and purpose. Scientists feared that — given the mission-orientation of these agencies — basic research would eventually fall victim to pressures for more immediate results. After three *years* of discussion and controversy on the organisation of research support, these concerns resulted in the formation of a government agency whose sole responsibility was the health of basic science and science education: the National Science Foundation (NSF).[5] With a budget for basic research that is a very small fraction of the budget spent on basic research by other agencies, NSF has concentrated on fields of science that are not otherwise well-supported and on strengthening science in universities.

---

4 In addition to the Radiation Laboratory at MIT, these included electronics research laboratories at Harvard and Columbia.

5 In addition to its own support of science, NSF was supposed to oversee and to co-ordinate the work of the other government agencies which were giving money for research. It did not, however, have the political or budgetary resources for meaningful co-ordination, and in 1963, its co-ordinating responsibilities were transferred to the Office of Science and Technology.

For largely historical and accidental reasons, then, we have ended up in the United States with a number of agencies supporting basic science, most of which have other primary goals.

Not only are there different agencies supporting science, there are different ways of going about this support. In general the pattern has been for certain agencies — notably the Atomic Energy Commission — to support large laboratories which have general goals and which use expensive tools such as accelerators and reactors. The Brookhaven and Argonne Laboratories, and the Oceanographic Institute at Woods Hole[6] are examples. Other agencies, in effect, set up small laboratories by supporting individual projects by individual scientists. This pattern has tended to produce debates about support of big science v. little science which are not in fact very meaningful. The sum total of small projects can be very large. Probably the bulk of research funds to universities — about $1 billion — goes for individual projects.

The tradition of individual project support has had important implications. It has greatly strengthened the role of individual scientists vis-a-vis their research institutions. It has also — until recently — avoided the politically sensitive issues of geographic distribution and institutional over-representation.

Two others methods of supporting basic science — used primarily by the National Science Foundation and the National Institutes for Health — are research fellowships for graduate students in science, university grants specifically for training scientists, and general institutional grants to universities doing scientific research. Such approaches accounted for about $350 million of last year's research budget, most of it in health-related research.

By and large, the money spent for basic science has gone to non-profit institutions — namely universities and government laboratories. This is not because industrial laboratories lack for scientific competence,[7] but because industry insists on controlling any

---

6 Woods Hole Oceanographic has recently affiliated with the Massachusetts Institute of Technology.

7 Starting even before World War II, large corporations have supported important basic research in areas in which they have a special interest. Irving Langmuir of General Electric and Clinton J. Davisson of Bell Telephone Laboratories won Nobel Prizes for work done in their respective industrial laboratories between World War I and World War II.

inventions which result from scientific work in their laboratories. The National Science Foundation supports university science almost exclusively, both because most of the best scientists are found there and because it furthers the NSF mission of science education.

The tough question that the government — or society — faces is not finding a mechanism for giving money for research but deciding how much money to spend where. Right after World War II, when research budgets were small, the government tended to rely on the scientific community almost completely to make these decisions. Politicians, despite their hardheaded and penny-pinching nature — bought the idea that only good scientists could determine what good science is. They did so for two simple reasons (in addition to the relatively small size of the total package under consideration): (1) when they asked natural scientists to justify their projects, they didn't understand the answers, and (2) programmes run by scientists had been a spectacular success during the war.

By 1960, however, the research budget even for basic science had become very large.[8] Politicians were becoming better staffed and more sophisticated in scientific matters. Not surprisingly, the politicians began to move in on the decision-making process. Having perceived a relationship between research and industrial growth, they began to ask, "Why doesn't *my* district get more contracts?" "Why should 15 universities get 50% of the nation's research funds?" Watching the total cost climb, they'd say "Would it be better to spend this money on applied research? on highways? on welfare? on housing?" "What are we *getting* for our money?"

Basic science is once again somewhat on the defensive. It is also closely involved in the political process. As an example of the present complexities of the role of government in science, I want to talk about high energy physics in the United States, particularly some of the decisions made by the government in the last five years.

---

8 In 1948, $100 million was spent on basic research; by 1956 it had risen to $350 million; and, in the aftermath of Sputnik, it rose to $800 million by 1960.

High Energy Physics: A Case Study of
Government in Science

High energy physics is the study of the ultimate constituents of matter and of forces and interactions between these constituents. In simplest terms, the smaller the size of the structure or interaction we wish to study, the greater the energy of the particle needed as a probe. When using a microscope, the smallest structure we can explore is determined by the wavelength and thus the energy of the light. To study structure on a still smaller scale, we have to use probes of shorter wavelength and higher energy — X-rays or gamma rays or even electrons. If we are to explore elementary particle structure, there is no alternative but to use probes of higher and higher energies. While important work has been done using cosmic rays, the great advances made in the field of high energy physics in recent years have been made by probing with particles that have been "artificially" accelerated.

Because high energy physics is so fundamental, it is very interesting to scientists and has attracted to its study some of the best scientists in the world. For the same reason, there is a great deal of prestige attached to making important discoveries in the field. If experiments in high energy physics were inexpensive, there is no doubt that governments — at least since World War II — would gladly support them. Particle accelerators, however, are expensive, and as the field has matured, the accelerators needed to advance our understanding have become more and more expensive because of their higher energies. Over time, these cost increases have changed the politics of support.[9]

From the end of World War II until the mid-1950s, there was considerable government support of research in high energy physics, but it received little attention from either the general public or the President.

After the war, both the Navy and the Atomic Energy Commission gave money for a number of particle accelerators, mainly cyclotrons, in laboratories all over the United States. These acceler-

---

9 For more detailed accounts of U.S. Government support of "big science", see Greenberg, *op. cit.*, pages 209-270, and David Z. Robinson, "Allocation in High Energy Physics", in Harold Orlans, ed., Science Policy and the University, Brookings Institution, 1968, pages 165-189.

ators were not very expensive, however, particularly in comparison to the large budgets of the military and the Atomic Energy Commission at the time.

The field advanced rapidly and scientists wanted to go to greater and greater energy levels. By using the principle of the synchrotron, invented by McMillan in the U.S. and Veksler in the Soviet Union, it became possible to get energies in the range above one billion electron volts (Bev).

Brookhaven National Laboratory and the Radiation Laboratory in Berkeley — both supported by the Atomic Energy Commission — began to construct synchrotrons (3 and 6 Bev and about $3 and $6 million respectively) about 1950. Research at these accelerators soon dominated the field. The decision to build and support these machines was considered to be a technical one and government policymakers at the highest level were not involved.

Then, in 1957, Stanford University came forth with a proposal for an electron linear accelerator two miles long that would produce electrons with energies of 20 Bev. The machine would cost 100 million dollars to construct and perhaps 25 million dollars a year to run. This was a highly technical proposal in an extremely esoteric field. But the scale of costs was one to which even the President of the United States had to pay attention.

Suppose you were the President or the Director of the Budget Bureau, would you approve the investment of $100 million in a two-mile drag strip for electrons? How do you decide?

Even in their technical aspects, decisions of this nature involve more than finding answers to the straightforward questions: "Is this machine a good design?" or "Will this machine significantly advance the field of physics?" The more difficult questions are relative: "What are the expected gains relative to the national investment of money and technical manpower?" "How does the importance of this machine compare with that of others that might come along later?" "Is spending the money this way more important than spending it for other high energy experiments? for some other area of physics? for some other area of science?"

In the end the real question is money (which, of course, translates into people and time). If the cost of a linear accelerator had been $1 billion instead of $100 million, it would never have been proposed. If the cost had been $10 million instead of $100 million, it would have been supported without question.

In any case, a panel of distinguished scientists was set up to study the value of the proposed accelerator to the high energy physics programme and to make recommendations.[10] Their report — which recommended construction of the Stanford accelerator but pointed out that other activities should still be supported — was presented personally to President Eisenhower.

The President made the final decision to go ahead and he announced it himself on May 14, 1959, in a banquet speech. With this step, high energy physics entered the political as well as the scientific arena.

As if to confirm this change of status, Congress refused for two years to appropriate money for the accelerator. The Congressmen acted partly out of personal pique — President Eisenhower had made the decision without consulting them — and partly out of a serious concern over appropriate mechanisms for planning such large programmes.

Despite the delays, the Stanford accelerator has now been built and it is operating very successfully. Meanwhile, high energy physics had acquired high political visibility.

Stanford had scarcely broken ground when another accelerator proposal arrived at the Atomic Energy Commission.

The Midwestern Universities Research Association (MURA) had set up, in the mid-1950s, a special centre for the design and construction of accelerators. The centre was staffed by exceedingly talented individuals who developed an imaginative design (the fixed-field alternating-gradient synchrotron) which allowed the acceleration of a much greater number of protons than a conventional synchrotron of the same energy.

---

10 The scientific panel as a governmental advisory device has been used extensively in the United States, and much less extensively in other countries. These panels have usually turned out to present a very useful point of view, and the scientists who serve on them work very hard. Certain cynical observers have suggested that the government only asks scientific panels to ratify things which the government is going to do anyway, and that panel reports are essentially "window-dressing". While there is no denying that reports which coincide with the government's eventual policy decisions get more circulation than those which do not, the evidence in no way supports the charge that panel findings have been fixed, predetermined, or subject to political influence.

By 1962, the MURA group prepared a detailed proposal for the construction of a high-intensity 10 Bev proton accelerator at a cost of 120 million dollars.   In addition to the proposal's scientific interest, it represented a way — probably the only way — to keep this remarkable design group together.

It was no secret that two other accelerator proposals were on the drawing board. A group at the Lawrence Radiation Laboratory in California was designing a 100 Bev accelerator,[11] and a group at Brookhaven was working on one in the 600-1000 Bev range.

Faced with this embarrassment of riches (or deficits?), the AEC and the President's Science Advisory Committee turned once more to a scientific panel.   The new panel — chaired by Dr. Norman Ramsey of Harvard — was asked to develop a ten-to-fifteen year plan for high energy physics (taking into account all three proposals), and to suggest priorities.

After eight months of intensive discussions and briefings, the Ramsey panel made a report. Its major conclusion was that the MURA accelerator was very valuable, but that the two other accelerators (which were some years away from design-readiness) were more important. The panel's recommendation was that "the Federal Government authorise the construction by MURA of a supercurrent accelerator without permitting this to delay the steps toward higher energy."[12]

The Atomic Energy Commission asked President Kennedy, in the fall of 1963, for permission to go ahead with the MURA accelerator. A decision was needed in December.

With the publication of the Ramsey panel report, the MURA proposal became the subject of intense political lobbying. Midwestern Congressmen (who felt very strongly that their area of the country had been shortchanged in the matter of government research money) mounted a general campaign of support. California Congressmen (armed with information supplied by some

---

11 Later the energy of the accelerator was changed to 200 Bev.

12 It should be noted that the democratic governments of this world find it very difficult to deal with recommendations of this sort. There just isn't enough control over future decisions. Who could really say whether the budget situation, the political temper, or the economic health of the country in five years would permit construction of another accelerator? Who could say whether construction would be delayed if another accelerator were built first?

scientists who feared construction of the MURA accelerator would delay construction of their own machine) argued against the MURA proposal. Partisans of the Argonne National Laboratory in Chicago (who thought the MURA machine would put Argonne's 12 Bev low-intensity accelerator out of business) helped to rally the opposition.

Shortly after taking office, President Johnson found himself facing not merely a technical and budgetary decision, but a political decision which had aroused Senators, Congressmen, and Chambers of Commerce all over the United States. "I devoted more personal time to this problem (the MURA accelerator)," President Johnson wrote later, "than to any nondefense question that came up during the budget process."[13]

Finally, despite a meeting with a delegation of scientists and Congressmen from the midwest, in January, 1964, President Johnson made his decision *not* to support construction of the MURA accelerator. Since he used the argument that a more important accelerator was in the offing, he committed himself, in effect, to the building of a 200 Bev accelerator at a later date.

Inasmuch as the Ramsey panel had recommended that the 200 Bev machine be built by the Lawrence Radiation Laboratory, most people assumed that the machine would be located somewhere in California. The President, however, asked the Atomic Energy Commission to search the country for the best possible site. Thus began the most recent — and most colourful — chapter in the politicising of the U.S. Government's relationship with basic science.

The 200 Bev accelerator is going to cost some $250 million. When complete, it will have an annual budget of $75 million and employ 5,000 people. The economic implications of these facts were not lost to the nation's livelier mayors, governors, senators, and university presidents. Every city or state that could produce 2,500 acres of flat open land, abundant water and electric power, and a modicum of transportation facilities promptly offered themselves to the AEC panel charged with selecting a site.

In the end, eighty-five long, detailed proposals were made to the AEC and they came from every state but North Dakota and

13 Letter to Senator Hubert H. Humphrey, January 16, 1964, quoted in Greenberg, *op. cit.*, page 267.

Massachusetts.   The state of Colorado spent thousands of dollars
on a slick-paper proposal with an inviting picture of a skier on
the cover.

Visiting teams of scientists were feted at every stop.   Scientists
found themselves at elaborate presentations with governors, mayors
and congressmen.   One state legislature went into special session to
pass a bill authorising the purchase of a site to donate to the
Federal Government for the accelerator.

After a great deal of detailed discussion, the AEC's site selection
panel recommended six possible sites.   Just *after* the 1966
Congressional elections, the Commission announced its choice — a
site in Weston, Illinois, near Chicago.[14]

Conclusion

We are in a period in which science will advance only through
the use of expensive tools.   Although high energy accelerators are
a conspicuous example of such tools, they exist in almost every
field.   Radio and optical telescopes cost millions of dollars on the
ground and even more if they're to be put in space, but they are
revolutionising our picture of the universe.   Significant weather
modification or prediction efforts require world-wide measurements
which are enormously expensive.   Advances in geophysics, earth
sciences and oceanography will take increasingly expensive research.
It took elaborate computation to make possible some of the
advances in molecular biology.   Chemists require computers,
counters, reactors and spectrometers.

Only the government can afford to underwrite these tools.   And
as the cost of basic research rises — both the aggregate cost of
supporting individual investigators and the cost of certain large-
scale experiments and expeditions — governments are going to ask
more questions.   If they are successfully to justify continued
research support, the scientific and university communities are
going to have to have a clear understanding of the value to society
of scientific activity.

---

14 The other finalists in the site sweepstakes were Sacramento, California;
   Denver, Colorado; Madison, Wisconsin; Ann Arbor, Michigan; and
   Brookhaven, L.I., New York.   Most of the Radiation Laboratory
   scientists declined to move to Weston.   Those who fought the MURA
   accelerator are probably contemplating the fact that had MURA's
   machine been built in 1962, political considerations would almost
   certainly have given the next large machine to California.

As costs rise, also, nonscientific factors play an increasing role in decision-making. Politicians are going to insist on taking account of possible technological fallout, applications to military and public health purposes, investment in underdeveloped regions of the country, geographic distribution of research funds, and the enhancement of international prestige.

I do not deplore this interaction of scientific and political factors. I welcome it. Science — even basic science — belongs in the real world. The ivory tower has become an anachronism, and both science and society will be the better for it.

As the marriage between science and government ripens into nagging middle-age, the research scientist is going to do a lot more explaining about his work. As a government official commented in a recent speech, "The relevance of science and research must be established anew with a sceptical Congress and with a generation that wasn't around when radar saved England from the Nazis". I am certain it can be done.

The Peaceful Uses of the Atom

(Five Chapters)

by

PAUL W. McDANIEL

P. W. McDANIEL,
Director of Research, U.S. Atomic Energy Commission,
Washington, D.C.

INTRODUCTION

Mankind is haunted by the problems of war, disease and physical and mental hardships. While knowledge of how to control nature has brought peace, health and happiness to some people in some parts of the world, there remains a dark spectre over all of us that meaningful and dignified lives cannot be enjoyed by a large number of inhabitants of the earth.

Atomic energy has enormous potential for destruction as evidenced by its initial use and by the reports of the past quarter of a century of major advances in its application to warfare.

But atomic energy offers much to society to help us reach our goals. It has often been said that the fear of nuclear war is a strong deterrent to war itself. This may be true but how much better the world would be if the basic underlying causes for conflict could be removed.

The atom offers mankind the unique opportunity, through cheap electrical and mechanical power, to free its citizens from covetous and greedy traits so common to those in fear of hunger, oppression and need.

In what follows in these chapters, I have endeavored to describe some of the peaceful uses which have been developed in the past twenty-five years for the atom. I have treated, for the most part, engineering and practical application to illustrate the enormous range of utility of the atom to man. Perhaps an even more practical application which threads through all of the atomic developments is the demonstration that through fundamental research free citizens are determined to remain free and to make it possible for a man to learn anything, especially those things that are *new* and *hard* and *deep*.

CHAPTER 1

The Development of Uses for Isotopes and Radiation

Introduction

More than a quarter of a century ago ways to release and use the energy locked within atomic nuclei were found. This discovery is one of mankind's most brilliant and decisive conquests of nature.

The conquest was brilliant and decisive not only because the individuals making the discovery showed unusual mental keenness in understanding the meaning of their research results, but also, because they were alert to their revolutionary consequences.

The Nature of Science

Like progress in almost all other fields of human endeavor, these early atomic energy developments, and those which followed, were made by the combined efforts of dedicated individuals from many countries. The laws of nature operate unchanged when national boundaries are crossed. Hence good scientific research takes no account of the nationality of its practitioners. Nature is blind not only to national boundaries but also to political philosophies, to religious faith, to wishful thinking or even to the urgent economic or social needs of mankind.

Those engaged in scientific research are compelled to think new thoughts, to be objective, truthful, accurate and patient in their search for some pattern in nature. The job of the scientist is not completed until he has communicated his findings, results, and methods to others, and until he has interpreted his results to his fellow citizens.

The search for truth in nature is a cooperative process. It consists of a series of steps, each one taken in the proper sequence and at the proper time. It is first necessary that the researcher be familiar with all existing and relevant knowledge — not with just some, but *all* relevant knowledge. He must have before him facts, laws and definitions. In order to make progress, he must examine critically all data before him. He may then create some tentative new idea or hypothesis which causally relates portions of existing knowledge. Using his tentative hypothesis and his general and specific knowledge, he may predict what would happen in nature if new conditions were imposed. Then the scientist can devise an experiment which will test the validity of his hypothesis. From the test results he may be able to derive new knowledge, from which new hypotheses, and their consequent usefulness in making further predictions can be tested. This is ever the cycle of science.

It has often been said that every step a scientist takes along the road to the solution of a problem must be taken initially in the footprints of his predecessors. But, when the scientist reaches that part of the path which no longer points the way to understanding, he must strike out in new directions, and with renewed efforts. He must hack out a new path from the brush and discover new ways to do things, or understand phenomena or explain nature. When he has done this he has the clear responsibility to outline the route he has taken so that his successors can readily know what he has done. If those who follow do not know what he has done, and how he did it, they cannot extend his work and achieve further understanding of the laws of the universe. Findings of scientists are thus of little permanent importance to mankind if their discoveries are not communicated to others.

The business of science, and of scientists, is thus to seek for the truth, whatever it be, and to discover as exactly as possible what things are and how they work. It is the business of society, and of each citizen, to determine the extent to which the laws of nature are used. Scientists, because of their understanding of nature, bear a special responsibility to society to assist in translating scientific findings to practical use.

Modern travel and communication techniques, two of the practical results of scientific discoveries, now make it possible for scientists and engineers to share each other's ideas and results on a current basis. Personal conversations between scientists have always been a vital link in the development of scientific thought. Not only do scientific minds stimulate each other but there are details of experiment, of theories, and of possible interrelationships which go unnoticed unless scientists can see and talk with each other personally. When this timely sharing is done, the chain of knowledge, necessary for deeper understanding of nature and its laws, can be forged quickly and effectively. When findings are not shared science and, as a consequence, society flounders.

Coulomb's Law

The events leading to the discovery of the law governing the forces between charged bodies are examples of how the progress of science is retarded when scientific results are not disseminated to others.

Henry Cavendish (1731-1810), a British scientist, is well-known for his brilliant work in chemistry and physics. The Cavendish Laboratories at Cambridge University serve as a reminder of this great scientist. But Cavendish would be better known and honored today had he published his discoveries in electricity. For more than a hundred years it had been known that two pieces of matter would repel each other if they had the same kind of electric charge. About 1800 Cavendish carried out many experiments on the forces which act between bodies bearing similar electrical charges. He proved to his own satisfaction that these forces obeyed the inverse-square law. When he had completed his work he simply let the matter drop and moved on to other experiments. He failed to publish his observations and his conclusions. He did not effectively communicate any of his results to his colleagues. As a consequence, almost fourteen years passed before a French scientist, Charles A. Coulomb, repeated the experiment and published his own discovery of the inverse-square law. "Coulomb's Law", as it is known today, is an important law describing the behaviour of charged particles. Indeed, it provides one of the keys to unlocking and releasing the energy from within atomic nuclei. There is no way to say just what turn science and society might have taken over the past two hundred years had Cavendish's discovery been published when he completed his experiments.

The Atomic Hypothesis

The idea that matter is not infinitely divisible but that all material bodies are composed of small, indivisible units called atoms, can be traced back some 2500 years to the early Greek philosophers and perhaps to Hindu scholars of an earlier age. The atomistic teachings of Leucippus and Democritus in the Fifth Century B.C., supported by Lucretius in the First Century B.C., made little headway, largely because of the objection of Aristotle. Thus, atomistic concepts lay dormant for several hundred years until they were revived during the Renaissance in Europe. Scholars of the 16th and 17th centuries such as Francis Bacon, Robert Boyle and Isaac Newton in England, Galileo Galilei in Italy, Rene Descartes in France, and others, favoured the view that matter was made up of ultimate particles or atoms.

By the end of the 18th century the work of the chemists of Europe and elsewhere had laid the basis for the development of the modern atomic hypothesis. Names like Dalton (British), Lavoisier (French), and Avogadro (Italian) attest to the universal character of science. It has been exactly one century since Mendeleev, the Russian scientist, first promulgated his periodic system of the elements.

In the closing years of the 19th century, it appeared that scientists were beginning to round out their basic understanding of nature and nothing more could be wanting. The doctrine of energy conservation and the laws of thermodynamics appeared to be capable of solving most of mankind's remaining problems. The mechanical and electrical laws of Newton and Maxwell explaining almost all known phenomena, had been formulated, tested and proven. Innovations had been made to such an extent in the social and economic life of the average man that many individuals, marvelling at man's ingenuity, felt that all worthwhile inventions must have already been made.

But in a short period of about ten years, from 1895 to 1905, several monumental scientific advances changed man's outlook on nature. He discovered new and then unexplainable phenomena. Phenomena which could not be understood in terms of the small, indestructible atoms governed completely by the mechanistic laws of Newton and Maxwell. These laws simply could not explain the phenomena of X-Rays and of radioactivity. The shock of this discovery shook the scientists of that time out of their complacent attitudes and did much to engender a new spirit of scientific investigation culminating in the great programmes of today.

Discovery of X-Rays, Radioactivity and Radium

Wilhelm Konrad Roentgen (a German experimentalist) in 1896 announced his discovery of a new highly penetrating radiation, which he called X-rays. These rays were shown to be of electromagnetic character. Within the next few months Antoine Henri Becquerel (a Frenchman) found that salts of uranium emitted radiations of a complex and puzzling nature. For some time the three types of radiation were known only as *Alpha, Beta* and *Gamm*a rays. Becquerel found that the alpha rays could be stopped by a thin sheet of paper, that the beta rays could penetrate

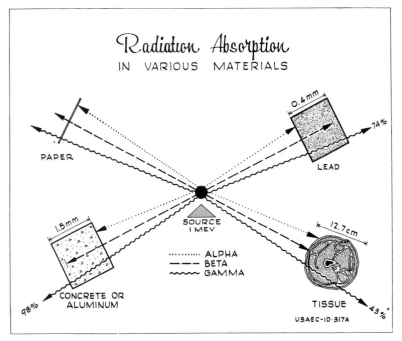

Figure 1-1. Radiation absorption in various materials.

several centimeters of a material like wood or glass, and that gamma rays were capable of penetrating several centimeters of dense materials like lead. He also found that these radiations could produce photographic images on sensitized plates, and that the alpha and beta rays were affected by electrical and magnetic fields. *Figure 1.1* illustrates the relative absorption of these radiations in materials. *Figure 1.2* illustrates the electrical properties of nuclear radiations.

The discovery of the radioactivity of uranium caused many individuals to turn their attention to this curious new phenomenon. Physicists, eager to learn the nature of these new radiations, carried out many new experiments. For example, the New Zealand physicist, Ernest Rutherford, then working with Frederick Soddy at McGill University in Montreal, is generally credited with being the first to identify the alpha ray components of the radiation emitted by uranium as being the charged nucleus of a helium atom. They also showed that the beta ray component was to be identified as an electron. The gamma ray component was identified as being of the electromagnetic type because of its similarity to X-rays.

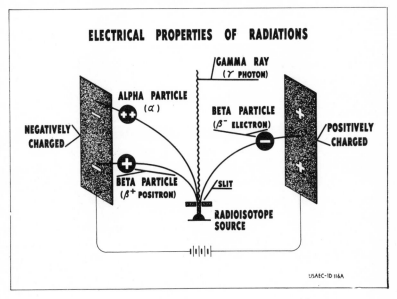

Figure 1-2. *Electrical properties of nuclear radiations.*

The husband-wife chemist team of Pierre Curie (French) and Marie Sklodovskei Curie (Polish) soon announced their discovery of radium. Throughout the world some of the best scientists devoted their attention to understanding the nature of radioactive decay and the laws of radioactive transformations. Others sought ways to apply these new materials to solving practical problems.

Within a few months of the discovery of X-rays the harmful effects of these rays on the skin were noticed. It was not long before radiation began to be used as a means for controlling the development of growth of cells, for example, in the treatment of some forms of cancer.

No less exciting than the important uses discovered for X-rays and radium in chemistry, biology, medicine and industry were the insights which were cast on the structure of the atom.

Rutherford's classical experiments on the bombardment of gold foils by alpha particles led him to conclude that the nucleus of the atom is small and dense and that it carried a strong positive electrical charge. Scientists further reasoned that the complete atom resembled a miniature solar system, with the sun corresponding to the nucleus and the planets to the electrons. This concept was highly developed by the great Danish physicist, Niels Bohr. The Bohr model of the atom is still widely taught because it explains many of the complex phenomena of nature from magnetism to optical spectra.

Transmutation

By the end of World War I, Rutherford and his colleagues were able to observe the transmutation of the element nitrogen into the element oxygen by bombarding the nitrogen with a beam of alpha particles. This experiment, reported almost exactly fifty years ago, was the first real demonstration that one element can be transformed into another element by nuclear bombardment. As simple as the equation is:

$$_2He^4 + {}_7N^{14} \rightarrow {}_8O^{17} + {}_1H^1 \tag{1}$$

it has such significant importance that it needs to be written here to serve as a reminder of the first man-made nuclear reaction. With the discovery in 1932 of the neutron, and of ways to accelerate charged atomic nuclei like the proton and the alpha particle to high energy, means have become available to initiate thousands of transmutation reactions in all of the chemical elements.

Discovery of the Neutron and Artificial Radioactivity

In 1930 W. Bothe and H. Becker in Germany reported that they had observed a highly penetrating radiation coming from beryllium, lithium and boron when these light elements were bombarded with alpha particles from a polonium source.

In 1932 James Chadwick in England succeeded in demonstrating that the results of Bothe and Becker could be explained by attributing the results to a new particle with about the same mass as the proton but with no electrical charge.

This particle, the neutron, within the decade following its discovery, was recognized as the key that made possible the release of atomic energy from the atomic nucleus. The discovery of the neutron and of the role it played in transmutations, provided scientists with the necessary information to permit a meaningful model of atomic nuclei to be developed.

During the course of a study in 1934 of the effects of alpha particles on the nuclei of some of the light elements, the French husband-wife team of F. Joliot and Irene Joliot Curie found that new radioactive forms of some of the elements had been formed. At about the same time it was discovered that neutrons could, by absorption into the nuclei of atoms, produce new radioactive forms of many of the elements studied.

Isotopes

Atoms that are chemically alike but which differ in mass are called **isotopes.** Isotopes have the same number of electrons orbiting about their nuclei and, therefore, have the same chemical characteristics. Isotopes have the same number of protons in their nuclei but they have different numbers of neutrons in their nuclei. We, therefore, say that isotopes have the same atomic number, Z, but different mass numbers, A.

Some combinations of protons and neutrons result in the formation of **stable** isotopes. Other combinations are unstable and nuclei containing these unbalanced combinations transform themselves into more stable nuclei by ejecting various particles or groups of particles. Energy is always released when these

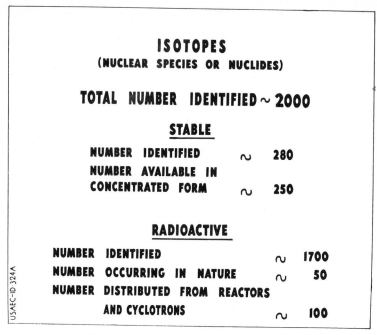

Figure 1-3. Number and kind of isotopes.

transformations take place.   An unstable isotope is called a
radioactive isotope or more commonly a **radioisotope**. Nearly 2000
stable and radioactive isotopes have been identified.   *Figure 1.3*
shows some statistics on the number and kinds of isotopes.

Radioisotopes decay with a rate characteristic of the individual
isotope.   The average time it takes for one-half of the amount of
a sample to disintegrate is called the half-life of that material.
Figure 1.4 illustrates the meaning of the term half-life.

Figure 1.5 is a chart which shows the radioactive and stable
isotopes for the elements up to Z = 8 (oxygen).   Note that
there is a general line of stable elements for those nuclei containing
equal numbers of protons and neutrons.   Nuclei containing more
neutrons than protons tend to be radioactive and to decay by
negative electron emission.   Those with more protons than
neutrons tend to disintegrate by positive electron emission
(positron).

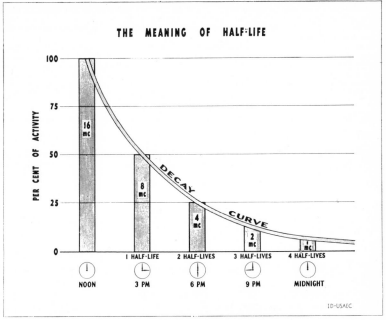

Figure 1-4. The meaning of half life.

The Isotopes of Hydrogen

Hydrogen is the lightest of all the elements. There are two stable forms of hydrogen found in nature. The mass of the heavier form (deuterium) is about twice that of the lighter form. The nucleus of the deuterium atom contains one proton and one neutron. Hydrogen occurs in nature as a mixture in the ratio of one part deuterium to about 6,600 parts of light hydrogen. The third form of hydrogen, called tritium, weighs almost exactly three times as much as the light isotope. Its nucleus contains two neutrons and one proton. Tritium is radioactive.

Figure 1.6 lists the three known isotopes of hydrogen, the lightest chemical element, and records some information about these isotopes. *Figure 1.7* illustrates some of these points.

When tritium decays a negative electron and a helium-3 nucleus results.

The equation is $_1\text{H}^3 \longrightarrow {}_2\text{He}^3 + {}_{-1}e^0 + \text{energy}$ (2)

The ejected electron can be detected because it is energetic enough to actuate electronic devices or to expose sensitive photographic film.

CHART OF THE ISOTOPES OF THE LIGHT ELEMENTS

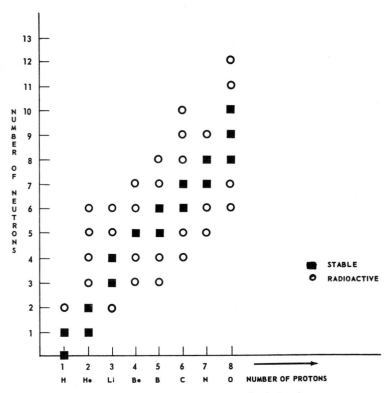

Figure 1-5. Chart of the isotopes of the light elements.

THE ISOTOPES OF HYDROGEN

Name	Hydrogen	Deuterium	Tritium
Symbol	$_1H^1$	$_1H^2$	$_1H^3$
	or P	or D	or T
No. Electrons	1	1	1
No. Protons	1	1	1
No. Neutrons	0	1	2
Abundance in Nature	99.985%	0.015%	Trace
Isotopic Mass (in atomic mass units)	1.007825	2.014103	3.016049
Half-life	Stable	Stable	12.26 years

Figure 1-6. The isotopes of hydrogen.

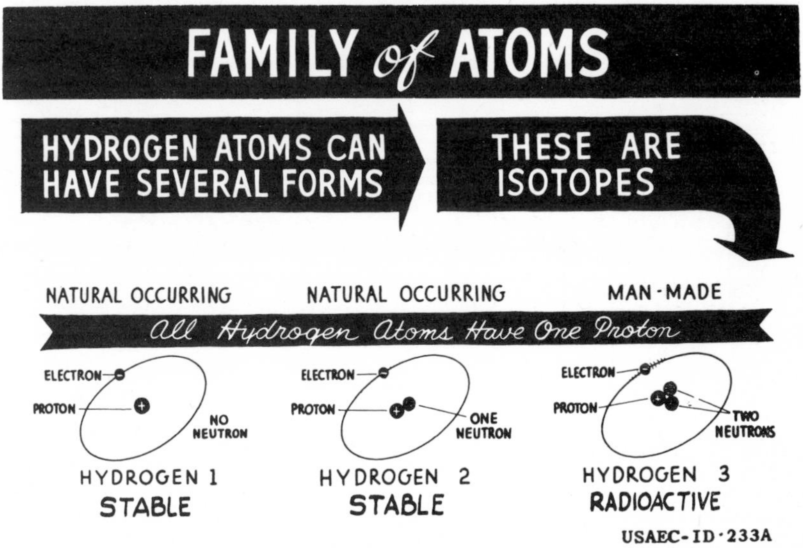

FAMILY of ATOMS

HYDROGEN ATOMS CAN HAVE SEVERAL FORMS

THESE ARE ISOTOPES

NATURAL OCCURRING   NATURAL OCCURRING   MAN-MADE

All Hydrogen Atoms Have One Proton

ELECTRON
PROTON
NO NEUTRON

ELECTRON
PROTON
ONE NEUTRON

ELECTRON
PROTON
TWO NEUTRONS

HYDROGEN 1
STABLE

HYDROGEN 2
STABLE

HYDROGEN 3
RADIOACTIVE

USAEC-ID·233A

Figure 1-7. The family of atoms.

Labelling with Isotopes

Since a radioisotope can be detected by the radiation which it emits, a technique called *labelling*, or *tagging*, becomes possible. It is, of course, possible in ordinary life to attach a label to a container describing either in words, or in some code, e.g., a flashing red or green light, the materials in the container. If the label remains attached, we know what is in the container.

So it is with isotopes. An element can be labelled by adding a small amount of one of its radioisotopes to it. Whatever happens to the element under study also happens to the radioisotope that was mixed with it. To locate the element even after it has undergone complete chemical or physical processes, one has only to look for the radiation from the radioisotope which was added. Frequently the element can be followed satisfactorily through successive operations if beakers containing filtrates, funnels with precipitates, or tubing through which a solution is flowing, are brought near a counter. *Figure 1.8* illustrates the sensitivity and selectivity of the isotope tracer method.

An element can also be labelled by the addition of a sufficient amount of one of its stable isotopes to significantly change the

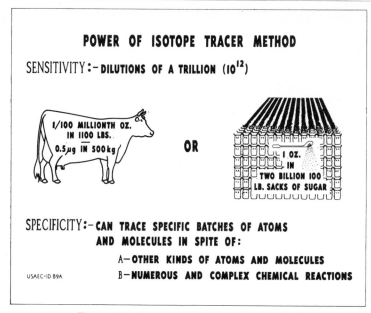

Figure 1-8. Power of isotope tracer method.

natural isotopic abundance of the element. The progress of the labelled element can then be followed by measuring the isotopic abundance of the various isotopes as the element moves through a series of physical and chemical processes. The use of radio-isotopes as tracers has had wider application than has the use of stable isotopes although there are many situations when only stable isotopes can be used.

Application of Isotopes and Radiation

Many books have been written on the thousands of problems in science, medicine, agriculture, and industry which have been studied using radioisotopes or their radiation. No brief account such as this can do adequate justice to the growing importance of the use of radiation and radioisotopes. In fact their use is so general that almost every research and engineering laboratory of any size has means to use them in their specialized work.

With this caution in mind, let us now turn to some of the more interesting and more significant results which have been obtained in several fields of endeavour. The particular applications selected for discussion were chosen, not so much for the importance of the application, but rather because the techniques used have more general application.

Agricultural Application

Annually in the United States about $12 billion are lost in crop production due to insects, weeds and plant disease. Approximately $4 billion of this loss is due to the loss of production caused by insects, $5 billion to weeds and $3 billion to plant disease. Similar, if not greater, losses occur throughout the world.

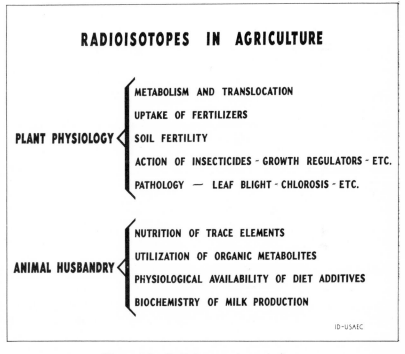

Figure 1-9. Radioisotopes in agriculture.

The use of radioisotope tracers and radiation in agriculture research has begun to reduce these losses. See *Figure 1.9* which lists some of the ways that radioisotopes are being used to study insects, plants, soil, and animals by agricultural experiment stations, universities, industry and the State and Federal governments.

Plant nutrition and metabolism

In order to properly fertilize crops one needs to know the mechanism by which plants absorb their nutrients from their environment. Radioisotopes are especially useful in these studies. By using radioactive phosphorus-32 in tracer experiments, scientists have found that a large percentage, perhaps as much as one-half to three-fourths of the phosphorus in a plant, was taken up in the first few weeks of the plant's growth. The application of these results is obvious because the results indicate that phosphorus fertilizers need only be used during the very early life of the plant.

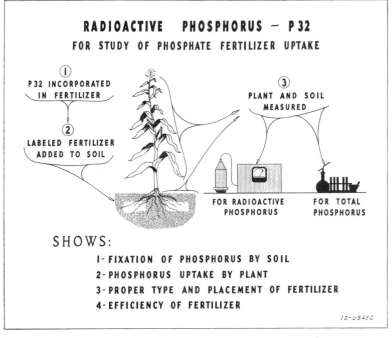

Figure 1-10. Use of P-32 in fertilizer uptake studies.

Figure 1.10 illustrates how radioisotopes are used in tracer studies of the phosphate uptake by plants from fertilizers. A small amount of phosphorus-32, a radioactive isotope of phosphorus, when incorporated in a fertilizer labels the phosphorus in the fertilizer when it is added to the soil. The amount of phosphorus from the

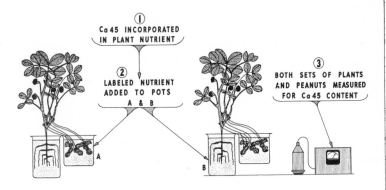

Figure 1-11. Use of Ca-45 in plant nutrition studies.

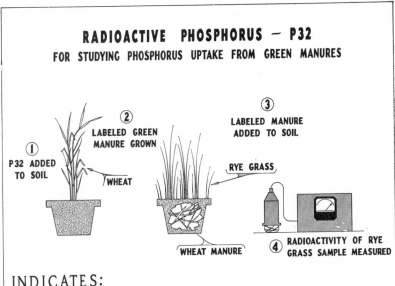

Figure 1-12. Use of P-32 in phosphorus uptake from green manure.

fertilizer absorbed by the living plant under various conditions of fertilizer application, can be measured by measuring the amount of radiophosphorus absorbed by the plant.

Figure 1.11 illustrates how calcium-45, a radioactive isotope of calcium, was used to study calcium nutrition of the growing peanut plant.

Figure 1.12 illustrates an experiment which used phosphorus-32 to study the phosphorus uptake by a growing plant whose roots were feeding on green manure. In this experiment phosphorus-32 was added to the soil nourishing growing wheat plants. After the

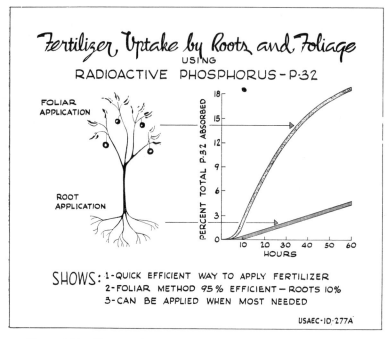

Figure 1-13. Use of P-32 in root and foliage fertilizer uptake studies.

wheat plants had grown sufficiently the green parts of the wheat plant were transferred into green manure to serve as the source of phosphorus for growing rye grass. The uptake of phosphorus from the green manure by the rye grass was studied by detecting the radiation from phosphorus-32 in the rye plant.

Figure 1.13 shows how the use of phosphorus-32 as a tracer answered questions of the relative amount of phosphorus uptake from fertilizer application to the roots and to the leaves of a plant.

Photosynthesis

Probably the most important process which takes place in green plants is photosynthesis. In these reactions the energy of the sun is used to convert carbon dioxide, water and other nutrients into complex and valuable substances. Photosynthesis is the basis for the total supply of food for the animal kingdom. A better understanding of the process of photosynthesis will surely result in real benefits to mankind.

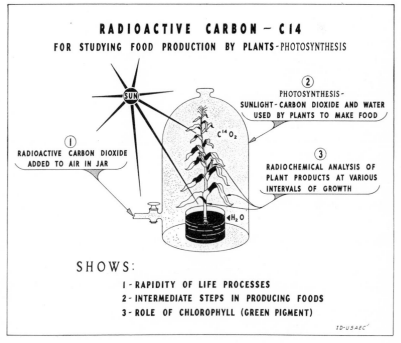

Figure 1-14. Use of C-14 in photosynthesis studies.

Figure 1.14 illustrates how radioisotopes are used in developing a further understanding of these complex reactions.

The chemical reactions leading to the formation in plants of the simplest sugars by the photosynthesis process are very complicated but may be represented by the equation:

$$6\ H_2O\ +\ 6\ CO_2\ \xrightarrow[\substack{\text{in presence}\\ \text{of sunlight}}]{\text{chlorophyll}}\ (CH_2O)_6\ +\ 6\ O_2\quad (3)$$

Here $(CH_2O)_6$ represents one of the sugars, glucose or fructose, found in growing plants.

Scientists had often asked whether the oxygen produced in photosynthesis originates in the carbon dioxide or in the water. In 1941 scientists at the University of California conducted experiments on reaction (3) with either the oxygen in the carbon dioxide, or the oxygen in the water, labelled with oxygen-18, a stable isotope of oxygen. The scientists found the oxygen-18 isotope in the product oxygen gas only if it were originally present in the water. They were thus able to show that the oxygen produced in photosynthesis originates from the oxygen contained in the water and not from the oxygen contained in the carbon dioxide.

More recent experiments by Professor Melvin Calvin and his co-workers at the University of California, using radioactive carbon-14 as the tracer isotope, have shed much light on the series of chemical reactions involved in the conversion of carbon dioxide to sugar, starches and celluloses. By combining the technique of paper chromatography and the incorporation of carbon-14, identifications of the compounds present could be made.

In paper chromatography a drop of liquid containing the mixture of chemicals to be separated is placed in a spot near the corner of a square piece of filter paper. By allowing a selected organic liquid to flow upward through the spot, the dissolved compounds are transported by capillary action to various locations in a column along the filter paper. By turning the filter through 90° and repeating the procedure with another solvent the chemical compounds in the original drop can be spread out in two directions. If this is done with different chemical compounds some of which contain carbon-14 and some of which do not, the position of the compounds that contain carbon-14 can be found by placing the filter paper in contact with a photographic film and making a radioautograph.

From such experiments, scientists have found that the first step in photosynthesis was the attachment of the carbon dioxide to some organic phosphate compound. Other similar experiments have shown that this material is ribulose disphosphate, a five carbon atom compound. These experiments also showed that the addition of a molecule of carbon dioxide to this compound splits it into two molecules of 3-phosphoglyceric acid. Further studies have established the principal reaction leading to the formation of the six carbon sugars. While much work remains before we can completely understand the process of photosynthesis, it is clear that the techniques of radioisotope tracers are playing an important role in these studies.

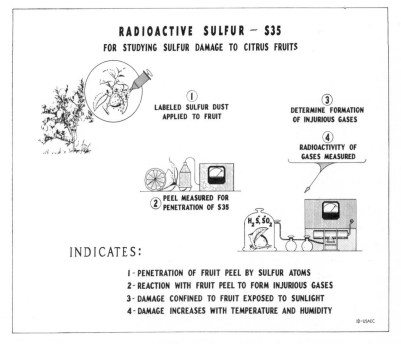

Figure 1-15. Use of S-35 in sulfur damage to citrus fruits.

Plant disease

There are thousands of organisms responsible for the plant diseases of the world. So far the best way found to reduce the economic losses from plant disease is to breed varieties of plants that are resistant to the particular disease. But nature provides not only for the development of disease resistant plant varieties but also for the changes in the disease organism itself. As a result, new strains of the disease organism attack the previously resistant plants. Great benefits might result if we knew more about the chemical treatment of plant disease. Radioisotopes are playing an important role in these studies.

For example, with radioisotope tracers it is possible to follow the effect of various fungicide chemicals when applied to plants. *Figure 1.15* illustrates one such application. Sulphur, used as a fungicide on citrus fruit, can penetrate the fruit peel and form gases which injure the fruit. Radioactive sulphur-35, used as a tracer, was able to help scientists determine ways to minimize the injurious effects as shown in the illustration. Much work continues to be done in the chemical control of plant disease and radioisotopes will be useful tools in these studies.

Weeds

The widespread use of chemicals to kill unwanted plants is well known. But all plants, even those we might wish to save, absorb these herbicides and suffer damage to their roots and shoots. In order to adopt those chemicals as herbicides which are most selective in the plants they kill, the use of radioisotope tracers should be most helpful.

Animal nutrition and metabolism

The food consumed by animals is either absorbed into the animal's body, retained for some time or excreted promptly. The basic problem of animal nutrition is to maximize the amount of food absorbed into the animal's body.

Radioisotope tracers have been used to determine the fate of nutrients fed to animals. For example, they have shown that 80 percent of the phosphorus contained in cows' milk comes from the cows' bones and only 20 percent from the feed. On the other hand, similar studies showed that only 35 percent of the phosphorus of eggs comes from the hen while 65 percent is provided by feed.

Radioisotopes can be used to study the role of trace amounts of various elements in diet deficiency nutritional problems. *Figure 1.16* illustrates this type of study.

RADIOACTIVE COBALT — CO 60
FOR STUDYING TRACE DEFICIENCIES IN DIET

CoCl₂

ADVANTAGES:

1 - RADIO-ANALYSIS RELIABLE FOR TRACE AMOUNTS
2 - ONLY TRACE AMOUNTS OF Co60 NEEDED
3 - METABOLIC ACTION INDICATED

ID-USAEC

Figure 1-16.   Use of Co-60 in studying trace deficiencies in diet.

Insects

Radioisotopes have been used to study the life cycles, dispersion, mating and feeding habits, and the parasites and predators of insects. For example, in Canada nearly a half million mosquito larvae were labelled with radioactive phosphorus-32. Some of the adults were found as far as seven miles away, but most were found within ⅛ mile of the place of birth.

The success in the use of radiation from radioisotopes to wholly eradicate the screwworm fly from the South-eastern United States is remarkable. The screwworm fly inhabits large areas of the world. The fly lays eggs in the open wounds of livestock and the burrowing maggots tear away living flesh and often kill untreated animals. In the South-eastern United States, damage from this insect amounted to $15 to $25 million annually. It had been suggested more than 30 years ago that these insect pests might be self-eradicated by rearing and releasing a large number of radiation sterilized males into the natural population.

In 1954, in a pilot experiment on the island of Curacao, male screwworm flies (made sterile by radiation treatment) were released at a weekly rate of 400 per square mile, three or four times the natural population. The insect was completely eradicated from the island.

Following this success in Curacao a campaign was begun in 1958 to attack the screwworm fly in the South-eastern part of the United States. More than two billion of these flies were hatched and grown to the pupal stage, irradiated with gamma rays to produce sterility, and permitted to mature. About 50 million of these adult sterile flies were released weekly over Florida, Georgia and Alabama. A fleet of 20 airplanes was used to handle the distribution. It had been hoped that the sterile males released in such large numbers would be fully competitive with the normal males of the region and that the number of matings thus resulting in sterile eggs would so outnumber the normal matings that the pest would be eliminated.

The plans of the entomologists were well rewarded. After a few weeks the number of eggs hatched from the normal native flies rapidly diminished to zero. After continuing the program for 18 months, the insect was completely eliminated.

The success of the programme inspired other workers in the United States and in other countries to try the method on other varieties of pests.

The oriental, melon, Mediterranean, and Mexican fruit flies; the pink bollworm and boll weevil; the sugar caneborer; the European corn borer; the gypsy moth; the codling moth; the tsetse fly and the anopheles mosquito are among the insect pests receiving continuing study.

While the rearing and release of sterile insects method is not the final answer for all major insect pests, it will continue to be given serious consideration where conditions needed for success are present.

Use of Isotopes and Radiation in Medicine

Radioisotopes have been used in many thousands of experiments in tracing the movements and distribution of important body substance in healthy and diseased individuals. They have proved to be particularly useful in the diagnosis and understanding of many diseases, especially where conventional diagnostic procedures have not been adequate. Stable isotopes are also useful in such studies.

Radiation has long been used to control the development and growth of cells, especially in the treatment of some forms of cancer. The use of the radiation from the element radium and from X-ray machines for therapy began more than fifty years ago. Since about 1950 the penetrating radiation from high energy accelerators has been used for the same purpose. The radiation from radioisotopes have proven to be especially convenient and effective in some therapeutic applications and large teletherapy units are widely available in hospitals and clinics throughout the world.

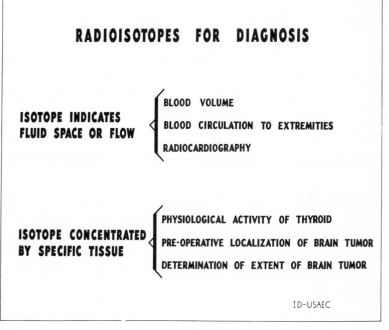

Figure 1-17. Use of radioisotopes in diagnosis.

The variety and extent of employment of isotopes in medicine both in the clinic and in the laboratories is indeed large.

Diagnosis

Modern medicine is highly dependent on diagnostic study techniques for the identification of disease and for evaluation of the progress of the patient after administration of treatment. Radioisotopes are often used by the diagnostician to aid him in his work. *Figure 1.17* indicates the wide range of the use of isotopes in diagnosis.

By giving a patient a minute amount of a selected radioisotope and by carefully following the fate of the radioisotope in the body, the physician can measure such things as blood flow, blood volume, digestion, metabolism, excretion and other bodily functions rapidly and accurately. *Figure 1.18* lists many of the uses made of tracer

ISOTOPES FOR TRACER STUDIES IN CLINICAL RESEARCH

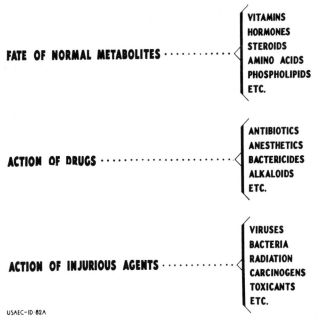

FATE OF NORMAL METABOLITES · · · · · · · · · · · ·
- VITAMINS
- HORMONES
- STEROIDS
- AMINO ACIDS
- PHOSPHOLIPIDS
- ETC.

ACTION OF DRUGS ·
- ANTIBIOTICS
- ANESTHETICS
- BACTERICIDES
- ALKALOIDS
- ETC.

ACTION OF INJURIOUS AGENTS · · · · · · · · · · · ·
- VIRUSES
- BACTERIA
- RADIATION
- CARCINOGENS
- TOXICANTS
- ETC.

USAEC-ID-82A

Figure 1-18. Use of radioisotopes in clinical research.

amounts of radioisotopes in studying the action and ultimate fate of drugs, injurious agents and normal metabolites. Some examples of the application of radioisotopes in the clinic will be reviewed to illustrate the range of possible uses.

Determination of water content of living body

Physicians often need to determine the total amount of water contained in the body of a patient under study. Now in the chemical laboratory it is an easy procedure to determine the water content on a lifeless sample of clay from an engineer's test boring. One way to do this is to weigh the sample before and after heating the sample to evaporate the water. The difference in weight is the weight of the water. But on a living patient such a drastic procedure is not possible. Isotopes, both stable and radioactive ones, have been used to make this determination.

About 35 years ago heavy water (D_2O) was first used to measure the total water content of a human body. The patient was given a known amount of heavy water by mouth or injection and enough time was allowed for it to mix uniformly with all the water in the body. A sample of the mixed water was then taken from the patient and analyzed for its heavy water content. The procedure was quite useful, but analysis of low concentrations of heavy water in ordinary water is difficult and the instrument used, the mass spectrometer, is complicated and costly.

When tritium became available, a new and more convenient method for measuring total body water was developed. This involved giving the patient an extremely small quantity of tritiated water (T_2O) and allowing this to mix uniformly with the body water. A sample of withdrawn body fluid was then tested for the quantity of radiation which it emits. The total water content of the patient's body can then be computed from consideration of the dilutions involved. *Figure 1.19* shows a technician carrying out this simple determination.

A somewhat similar procedure can be used to determine the volume of the blood plasma in a living body. *Figure 1.20* shows a schematic outline of radioiodine-131 being used for this purpose.

Studies of the thyroid gland

The functions of the thyroid gland have been intensively explored with radioisotope techniques. The convenient location of the

Figure 1-19. The determination of total body water using tritiated water.

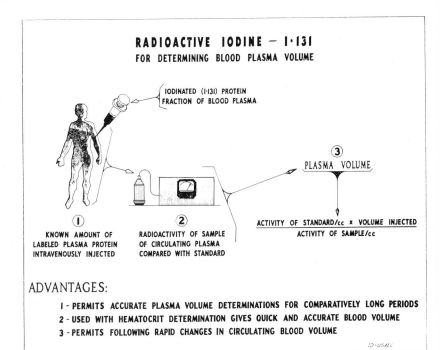

Figure 1-20. Use of I-131 in blood plasma volume determination.

thyroid gland near the surface of the neck, the extraordinary ability of this organ to concentrate the element iodine, and the ready availability of the iodine-131 isotope, has made it possible to investigate the uptake pattern, the retention and release of iodine by the thyroid gland, and the protein-bound fractions in the circulating blood. *Figure 1.21* shows some of the techniques used in studying thyroid gland physiology. In these studies the patient drinks a small amount of Iodine-131. The thyroid gland takes up most of the radioiodine retained by the body. A detector measures the amount of radiation emitted by the absorbed radio-iodine allowing the physicians to understand the physiological activity of the gland.

Figure 1.22 illustrates a simple procedure for determining hyper-thyroidism, a common condition, in human patients. In this test the individual is given a very small amount of radioactive sodium iodide to drink and about two hours later the radiation which comes from the neck area of the patient is measured. The

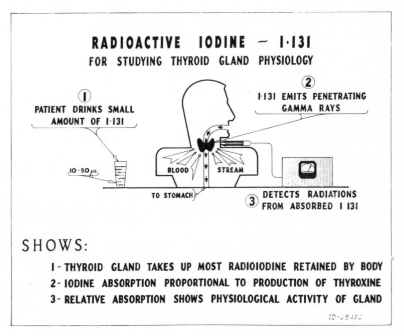

RADIOACTIVE IODINE — I·131
FOR STUDYING THYROID GLAND PHYSIOLOGY

① PATIENT DRINKS SMALL AMOUNT OF I·131

② I·131 EMITS PENETRATING GAMMA RAYS

10·50 μc

BLOOD STREAM

TO STOMACH

③ DETECTS RADIATIONS FROM ABSORBED I 131

SHOWS:

1- THYROID GLAND TAKES UP MOST RADIOIODINE RETAINED BY BODY
2- IODINE ABSORPTION PROPORTIONAL TO PRODUCTION OF THYROXINE
3- RELATIVE ABSORPTION SHOWS PHYSIOLOGICAL ACTIVITY OF GLAND

Figure 1-21. Use of I-131 in thyroid gland physiology studies.

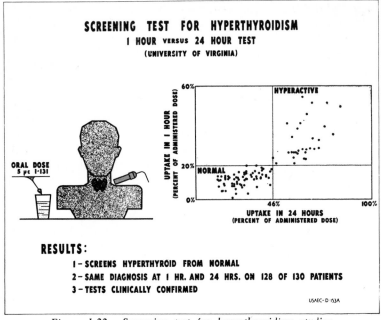

Figure 1-22. Screening test for hyperthyroidism studies.

uptake of radioactive iodine in the patients with hyperthyroidism varies dramatically from the uptake in normal individuals, as shown in the chart in *Figure 1.22.*

Blood circulation

It is often important to understand the condition of the blood circulation in a patient. A sample of salt solution labelled with sodium-24 can be injected into a vein in the arm or leg of an individual. As the blood circulates the sodium-24 tracer can be detected with a radiation counter as it passes another part of the body. The elapsed time and the magnitude of the reading is a good indication of the presence or absence of constriction or obstruction in the circulatory system. *Figure 1.23* illustrates this procedure.

Cancer studies

Malignant (cancerous) cells are characterized by the abnormally high rates at which they increase in number. Radioisotopes have

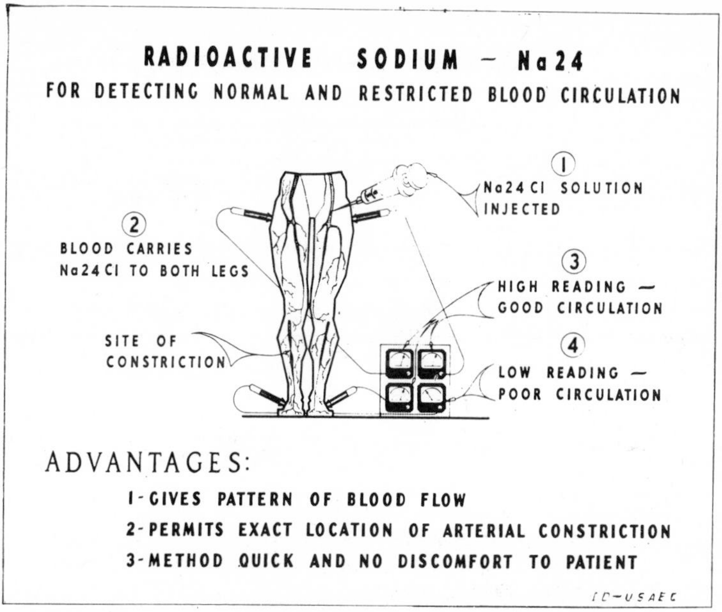

Figure 1-23. Use of Na-24 for detecting abnormal blood circulation.

CANCER STUDIES AIDED BY RADIOISOTOPES

- CITRIC ACID CYCLE
- NUCLEIC ACID METABOLISM
- ANTICANCER DRUGS
  - ENZYME INHIBITORS
  - NITROGEN MUSTARDS
  - METABOLIC SUBSTITUTES
- SYSTEMIC EFFECTS OF CANCER
  - FUNCTIONING OF OTHER TISSUES
  - ANEMIA
- CARCINOGENS
  - RADIATION
  - AMINOAZO DYES
  - HYDROCARBONS
- INTRACELLULAR RADIATION EFFECTS

USAEC-ID-234A

Figure 1-24. Use of radioisotopes in cancer studies.

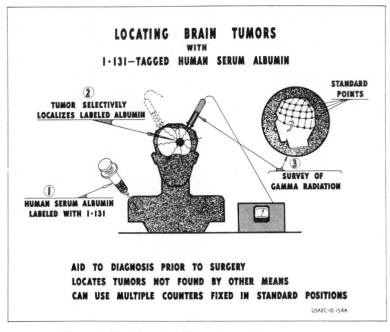

Figure 1-25. Use of I-131 to locate brain tumors.

been very useful in understanding the complex cellular processes which are manifested in cancer. *Figure 1.24* is a list of some of the cancer studies which have been aided by radioisotopes.

Location of brain tumor

Substances injected into the blood stream normally do not pass into the interior of brain cells. Brain tumors often alter this barrier between the blood and brain in such a way that albumin (the serum of the blood) can penetrate tumorous tissue. This suggests a way to locate a brain tumor by labelling the serum with radioactive iodine and later scanning the skull for hot spots of radiation. *Figure 1.25* illustrates this procedure.

Whole body scanning

Instruments have been developed that enable the physician to determine the distribution of radioisotopes within the living body. While these instruments, called whole body scanners, vary in their construction and operation, they all basically employ a collimated,

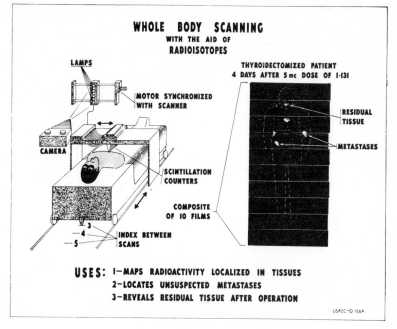

Figure 1-26. Whole body scanning.

movable radiation detector and some mechanism for translating the electrical impulse in the detector with a visual representation.

Figure 1.26 illustrates the principle of whole body scanning. This has proven to be a successful diagnostic technique because compounds labelled with radioisotopes may concentrate in different parts of the body. With the proper selection of labelled materials the physician can study metabolism in a wide variety of disease conditions. A study of such scans can often assist the physician in locating tumors, especially those which occur within the brain. Whole body scanning has been used to provide information necessary for performing needle biopsies to diagnose kidney or liver disease. Kidney scans to determine the size and shape of this organ have been helpful in treating persons with mild and moderate uremia.

Scanning techniques have been also useful in the diagnoses and treatment of heart disease, bone abnormalities and in many other ways.

Therapy

There are several ways to use the radiation from radioisotopes for therapeutic purposes. All of these methods have as a common principle the selective irradiation of some specific part of the body with a controlled amount of radiation and at the same time the irradiation of the other parts of the body with as little radiation as practical. *Figure 1.27* lists some of these methods.

Polycythemia vera

Polycythemia vera is a disease which was among the first to be treated about 30 years ago with artificial radioisotopes. This is a serious, chronic ailment characterized by an abnormal increase in total blood volume and in the number of red blood cells. A useful treatment, but by no means a cure, for this ailment, is to reduce the rate of formation of red blood cells by irradiation of the blood forming tissue of the body.

Radioactive phosphorus-32 emits beta rays that can penetrate several millimetres of tissue. When phosphorus, spiked with

THERAPY METHODS WITH RADIOISOTOPES

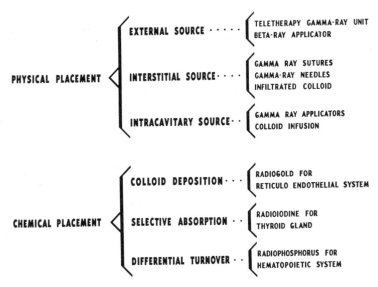

Figure 1-27. Therapy methods with radioisotopes.

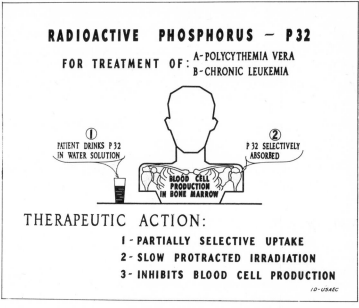

Figure 1-28. *Use of P-32 for treatment of polycythemia vera.*

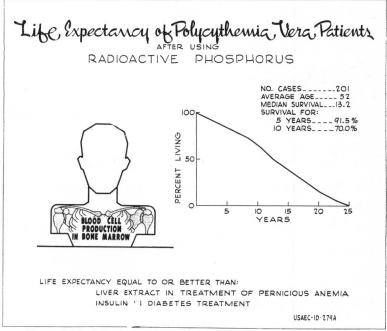

Figure 1-29. *Life expectancy of polycythemia vera patients treated with P-32.*

phosphorus-32, is absorbed in the body a great deal of it is selectively concentrated in the blood forming hematopoietic cells in the marrow of the bones. Here the beta rays from the decaying phosphorus-32 attack the blood forming tissues and inhibit the excessive formation of red blood cells thereby relieving the abnormal situation.

Figure 1.28 illustrates the application of this therapeutic treatment. This disease can be so successfully controlled with this technique that life expectancy is now about the same as that for diabetic patients treated with insulin, or pernicious anemia patients treated with Vitamin B_{12}. *Figure 1.29* shows the survival of 201 cases of polycythemia vera patients after phosphorus-32 therapy.

Particle beams from accelerators

Soon after the large machines for the acceleration of charged particles were invented their special value to therapy was recognized. It was observed that under certain conditions a monoenergetic beam of protons or alpha particles could be stopped abruptly in tissue with an effect similar to a surgeon's knife. Thus it became possible to deliver radiation energy deep within the body to carefully localized sites without undue damage to nearby tissue. After the successful treatment in 1958 of some kinds of cancer in experimental animals, a beam of high energy protons was used to destroy the pituitary gland of a patient with advanced cancer of the breast. Following this successful treatment, many hundreds of patients have received treatment for metastatic breast carcinoma, acromegaly, Cushings disease, brain tumors and other serious abnormalities. Very gratifying results have been obtained in most of these patients and it is clear that heavy particle beams from accelerators will be used increasingly in future years for the treatment of these and other diseases.

Radioisotope needle implantation

Small capsules or needles containing selected radioisotopes can be implanted surgically within a particular organ and that organ subjected to the radiation emitted by the radioisotope. Cobalt-60, gold-198 and yttrium-90 are commonly used for this purpose. Yttrium-90 being a pure beta emitter, is especially useful in those cases where local action is required, for example in the partial removal of the pituitary gland.

RADIOISOTOPE - POWER CARDIAC PACEMAKER

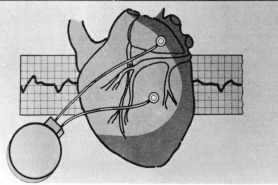

Figure 1-30. *Radioisotope-powered cardiac pacemaker implantation principle.*

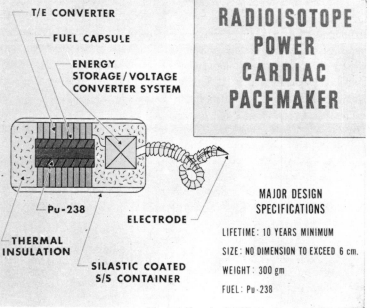

RADIOISOTOPE POWER CARDIAC PACEMAKER

T/E CONVERTER

FUEL CAPSULE

ENERGY STORAGE / VOLTAGE CONVERTER SYSTEM

Pu-238

ELECTRODE

THERMAL INSULATION

SILASTIC COATED S/S CONTAINER

MAJOR DESIGN SPECIFICATIONS

LIFETIME: 10 YEARS MINIMUM

SIZE: NO DIMENSION TO EXCEED 6 cm.

WEIGHT: 300 gm

FUEL: Pu-238

Figure 1-31. *Radioisotope-powered cardiac pacemaker major design specifications.*

Heart Reinforcement Devices

Heart disease claims about 700,000 lives in the United States annually. A recent study by the National Heart Institute indicates that about 100,000 of these might be helped if circulatory support systems were available for permanent insertion into the human body.

Pacemakers

One such device relies on the use of regular pulses of electricity to shock a defective heart into a normal beating cycle. Battery powered Pacemakers, as these heart assist devices are known, are occasionally used today. However, battery replacement is obviously a major problem.

Work is in progress to develop a radioisotope-powered Cardiac Pacemaker. *Figure 1.30* illustrates how the Pacemaker would be used. The entire device would be surgically implanted and hopefully would have a minimum life span of 10 years.

Figure 1.31 lists the major specifications and the general schematic design for a radioisotope-powered Cardiac Pacemaker. The device would be fuelled by converting the heat from the decay of plutonium-238 into electricity by the thermoelectric effect. A total of $0·135$ thermal watts of isotopic power appears to be sufficient to power the device. *Figure 1.32* is a photo showing parts of an encapsulated plutonium-238 heat source for the unit. *Figure 1.33* is a photo showing the size of a prototype electronics circuit in relation to a U.S. dime and *Figure 1.34* is a sketch showing the major components of a nuclear Pacemaker.

Artificial heart

The replacement of a diseased human heart by a healthy heart from a recently deceased person has become a reality. Many such transplants have been carried out with varying success. Current studies indicate that radiation can play an important role in counteracting the immunity reaction which inhibits transplanting of natural organs.

In the not too distant future the replacement or reinforcement of the human heart by a mechanical pump, implanted within the body, may become a practical reality. Such a pump would require a long-lived, implantable power source to drive it. The possible use of a radioisotopic heat source for this purpose is under active

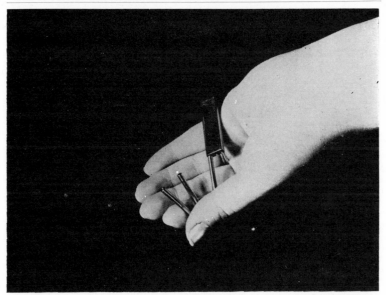

Figure 1-32. Radioisotope-powered cardiac pacemaker Pu-238 heat source.

Figure 1-33. Radioisotope-powered cardiac pacemaker electronic circuit.

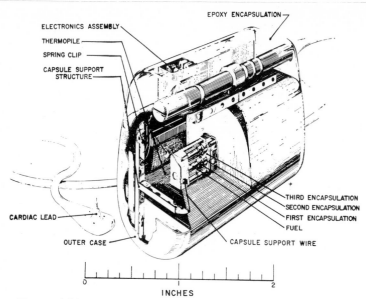

ELECTRONICS ASSEMBLY
THERMOPILE
SPRING CLIP
CAPSULE SUPPORT STRUCTURE
EPOXY ENCAPSULATION
THIRD ENCAPSULATION
SECOND ENCAPSULATION
CARDIAC LEAD
FIRST ENCAPSULATION
FUEL
OUTER CASE
CAPSULE SUPPORT WIRE
INCHES

Figure 1-34. Radioisotope-powered cardiac pacemaker major components.

investigation. Problems under investigation include (1) development of the radioisotopic heat source, (2) safety evaluation of using radioisotopic sources in the wide scale treatment of heart disease, (3) radiobiological studies, (4) development of materials compatible with blood, (5) development of pumps, and (6) evaluation of the effect of additional heat sources within the human body.

Results to date indicate that a radioisotopic fuelled implantable power source is feasible. *Figure 1.35* shows in general how the various components of the artificial heart might be implanted surgically within a human body.

If these developments succeed useful years might be added to many lives.

Industrial Application of Radiation and Radioisotopes

Industry has begun the large-scale application of radiation from radioisotopes and from electronuclear machines to the national goals, economy and welfare.

The radioisotope and radiation industry in the United States is

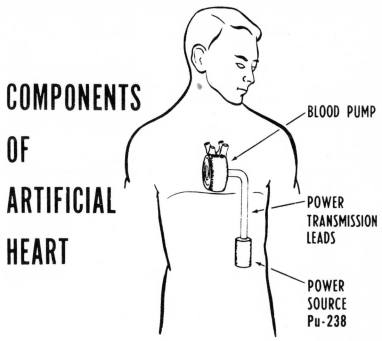

COMPONENTS

OF

ARTIFICIAL

HEART

BLOOD PUMP

POWER
TRANSMISSION
LEADS

POWER
SOURCE
Pu-238

Figure 1-35. Components of artificial heart.

characterized by significant participation by private companies and by rapid annual growth. These organizations are involved in the major aspects of radioisotope production and distribution including the private production of radioisotopes and commercial sale of radiochemicals, radiopharmaceuticals, radiation sources, and nuclear devices such as gauges.

The volume of such commercial business is more than $50 million a year and growing rapidly. For example, about 20 companies are engaged in distributing about $8-$10 million annually of radiochemicals. Radiopharmaceutical sales total about $15 million annually. These materials are used each year in the diagnosis of between 3 and 4 million patients. Radioisotopes for use as radiation sources for process radiation application, medical teletherapy, industrial radiography and process control and similar applications account for $3-$5 million annually. Radioisotope gauging has achieved widespread acceptance by U.S. industry,

RADIOISOTOPES IN INDUSTRY

FIXED SOURCE [MEASURE CHANGE IN RADIATION INTENSITY] · · · · {
RADIOGRAPHY
THICKNESS GAGE
LIQUID LEVEL GAGE
DENSITY METER

MOVABLE SOURCE [LOCATE OR FOLLOW MARKED OBJECT] · · · {
LIQUID FLOW THROUGH PIPE
LOCATION OF "GO DEVIL"

TRACER [PHYSICAL TRANSFER] · · · · · · · · · · {
FRICTION WEAR
SOLID DIFFUSION

TRACER [PHYSICAL - CHEMICAL TRANSFER] · · · {
DETERGENCY
MINERAL FLOTATION
MOVEMENT OF PRESERVATIVE

TRACER [MECHANISM OF REACTION] · · · · · · {
ROLE OF CATALYSTS
FISCHER - TROPSCH SYNTHESIS
SOURCE OF COKE SULFUR

USAEC-ID-427

Figure 1-36. Radioisotopes in industry.

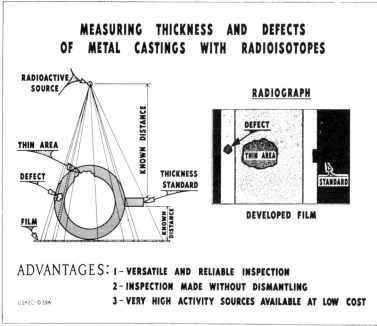

MEASURING THICKNESS AND DEFECTS
OF METAL CASTINGS WITH RADIOISOTOPES

RADIOACTIVE SOURCE

RADIOGRAPH

KNOWN DISTANCE

THIN AREA

DEFECT

DEFECT

THIN AREA

THICKNESS STANDARD

STANDARD

FILM

KNOWN DISTANCE

DEVELOPED FILM

ADVANTAGES: 1 - VERSATILE AND RELIABLE INSPECTION
2 - INSPECTION MADE WITHOUT DISMANTLING
3 - VERY HIGH ACTIVITY SOURCES AVAILABLE AT LOW COST

USAEC-ID 59A

Figure 1-37. Radiography of metal castings.

where 5,000 or more such devices are in current routine use. This market is served by several firms and total sales are about $25 million annually.

Radioisotopes in industry

Radioisotopes have found so many uses in industry that it becomes difficult to quote meaningful statistics concerning the various applications. Thickness and level measurements probably account for about one-fourth of all radioisotope application in industry. Gamma ray radiography probably accounts for another one-fourth of the total. The remaining one-half is divided among tracer applications, activation analysis, control of equipment, locating objects, luminous paints and miscellaneous applications. *Figure 1.36* is a chart showing some of the uses of radioisotopes in industry.

Radiography

Radioisotopes, particularly those with relatively long half lives and emitting gamma rays, have found widespread use in measuring the thickness of dense, opaque bodies and in determining the location of hidden defects therein. The techniques used are simple and similar to the well known techniques of taking X-rays. *Figure 1.37* illustrates how such radiographs are made. *Figure 1.38* lists the characteristics of various isotopes used in industrial radiography. The use of radioisotopes in radiography requires detailed consideration of many pragmatic factors such as optical arrangements, photographic emulsion processing techniques and film sensitivity.

Thickness and level measurements

Radioisotope gauging techniques are now routine for many industries in the United States. The quantities being measured may be static, such as the thickness of a coating of paint or the level of the contents of a tank, or they may be dynamic, such as the movement of products on a conveyor belt or the rolling of paper, plastic or metal into a sheet. *Figure 1.39* shows a sketch of the use of a radioisotope to measure and control the thickness of a product of a rolling operation.

Beta ray gauges are the most widely used in thickness and level measurements because the penetrating power of beta rays is more

ISOTOPES FOR RADIOGRAPHY

ISOTOPE	HALF-LIFE	PRINCIPAL GAMMA RAY ENERGIES—Mev	APPROX. SPEC. ACT.	APPROX. γ-RAY OUTPUT/CURIE r/hr/ft	TYPE OF USE
Co 60	5.3 y	1.1, 1.3	7 c/g — NOW 35 c/g — 1954	14.4	HEAVY CASTINGS THICK WELDMENTS
Ta 182	117 d	1.1, 1.2	1.5 c/g GREATER IF NEEDED	6.7	HEAVY CASTINGS THICK WELDMENTS
Cs 137	37 y	0.66	20 c/g — 1955 (Cs_2SO_4)	3.5	—
Ir 192	75 d	0.3, 0.5	2.5 c/g — NOW 30 c/g — 1954	3.0	PIPE LINES LIGHT ALLOYS THIN SECTIONS
Tm 170	127 d	0.085	SERVICE IRRADIATION	—	LIGHT ALLOYS THIN SECTIONS

USAEC-ID-149A

Figure 1-38. Isotopes for radiography.

RADIOACTIVE SOURCE
FOR GAGING THICKNESS

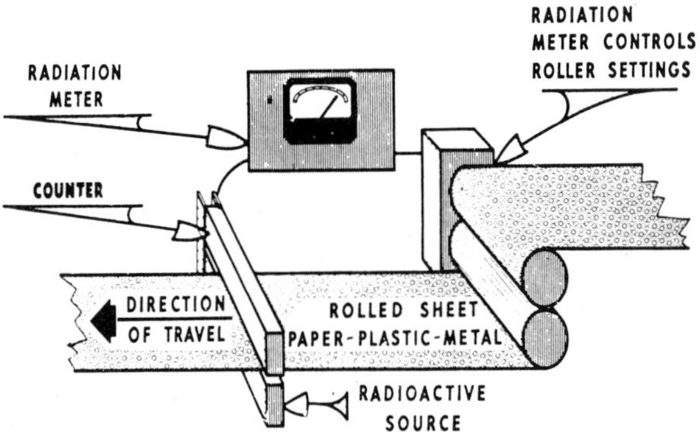

RADIATION METER

RADIATION METER CONTROLS ROLLER SETTINGS

COUNTER

DIRECTION OF TRAVEL

ROLLED SHEET PAPER-PLASTIC-METAL

RADIOACTIVE SOURCE

ADVANTAGES:

1 - RADIATION SOURCE SELECTED TO SUIT MATERIAL
2 - NO CONTACT - NO TEARING - NO MARKING MATERIAL
3 - RAPID AND RELIABLE

USAEC-ID-69A

Figure 1-39. Radioisotopes thickness gauge.

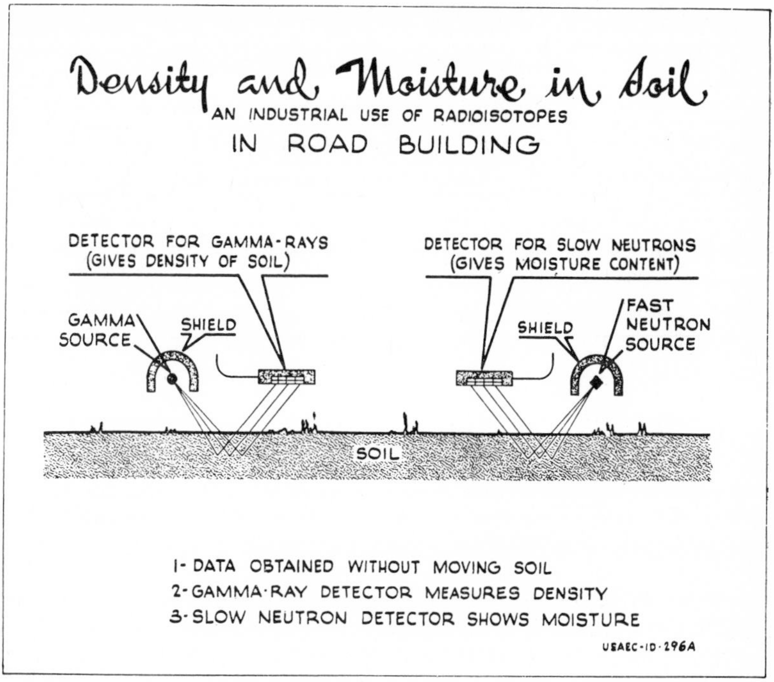

Density and Moisture in Soil
AN INDUSTRIAL USE OF RADIOISOTOPES
IN ROAD BUILDING

DETECTOR FOR GAMMA-RAYS
(GIVES DENSITY OF SOIL)

DETECTOR FOR SLOW NEUTRONS
(GIVES MOISTURE CONTENT)

GAMMA SOURCE   SHIELD

SHIELD   FAST NEUTRON SOURCE

SOIL

1- DATA OBTAINED WITHOUT MOVING SOIL
2- GAMMA-RAY DETECTOR MEASURES DENSITY
3- SLOW NEUTRON DETECTOR SHOWS MOISTURE

USAEC-ID-296A

Figure 1-40. Use of radioisotopes in soil density and moisture determinations.

compatible with most of the materials being measured than are alpha ray or gamma ray emitters.

Methods have been adopted for measuring the thickness of items such as: corrosion in a vat when the interior is inaccessible, the carbon deposit in the cylinder of internal combustion engines, a concrete pipe in oil bore holes, electrodeposited platings, coal seams, and hundreds of other items.

It is often important to be able to measure the density of the soil at a particular location, especially in road building. The usual method is to remove a sample of the soil and examine it in a laboratory. But this is time consuming and subject to some uncertainties. By using a radioisotope like cobalt-60, which emits gamma rays the density of the soil may be measured quite accurately by using the back-scattering of the gamma rays from the subsoil as shown in *Figure 1.40*. This figure also shows a similar method for measuring the moisture content of the soil by measuring the back-scattering of slow neutrons from the water in the subsoil.

Environmental Studies

Radioisotopes are finding uses in environmental studies, for example in tagging and tracing the sulfur dioxide from industrial smoke stacks emitting this pollutant into the atmosphere as well as in monitoring the amount of SO_2 passing through an operating stack. By using tracer techniques the flow rate, ion and particulate migration and tracing of the dispersion of pollution from individual contributors can be carried out effectively.

One can also use radioisotopes for the determination of beach erosion, sediment transport, for measuring current flow, for dating the water at great ocean depths, for *in situ* analysis of ocean deposits, and for many other factors of importance to oceanographers.

Transuranium Isotopes

The discovery and investigation of the elements beyond uranium in the periodic table has led to interesting and important new uses for radioisotopes. The enormous practical importance of plutonium-239, both as the explosive ingredient of nuclear bombs and as a fuel from which electricity can be generated, is well known. As we learn more about the isotopes of the newly discovered elements beyond plutonium in the periodic table, we are sure to find interesting and important applications.

Plutonium-238 and curium-244, both alpha particle emitters with $86 \cdot 4$ year and $18 \cdot 1$ year half lives respectively, have found wide uses in heat sources for space and terrestrial application. Mixtures of the radioactive isotopes of curium and americium are also finding use as heat sources in industry and elsewhere.

Californium-252 emits neutrons by the spontaneous fission process. It is such a potent source of neutrons that a minute quantity of californium-252 can be used in a variety of applications where large and costly reactors are now required. A program for the production of significant amounts of californium-252 and other transplutonium isotopes is under way in the United States. *Figure*

SCHEME OF THE NATIONAL TRANSPLUTONIUM
ELEMENT PRODUCTION PROGRAM

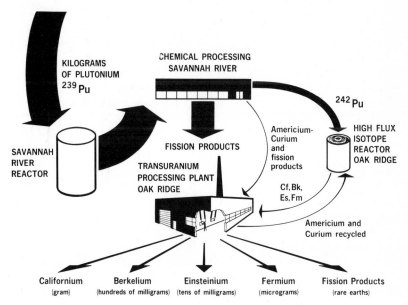

*Figure 1-41. Scheme of the national transplutonium element
production programme.*

1.41 shows the scheme we are now using for the preparation of
research amounts of these materials. As soon as we are able to
demonstrate requirements for quantities of californium-252, we
would plan to go into production on a larger scale. In these
processes, Plutonium-239 serves as the raw material for the
production of other transplutonium elements by successive
neutron captures. *Figure 1.42* illustrates the major path of such
productions.

As an example of the type of use to which californium-252 can
be put is shown in *Figure 1.43*. Here a tiny amount of this isotope
is used as a neutron source to irradiate a sample of low-grade gold
ore. The gold in the sample is made radioactive and the radiation
can then serve as a measure of the amount of gold present in the ore.

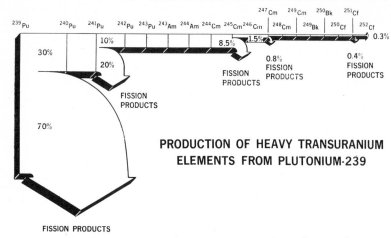

Figure 1-42. *Production of heavy transuranium element from plutonium-239.*

Nuclear archaeology

The atom has become useful to the archaeologist in determining the places and origin for pottery shards thereby helping him to solve the mystery of where ancient man travelled and what he did. The methods used by the nuclear chemists combines neutron activation analysis and gamma ray spectroscopy.

On the assumption that pottery clay from a particular location has a distinctive composition, nuclear chemists are able to compare unfamiliar pieces of pottery with shards whose origin is known. Pottery fragments irradiated in a nuclear reactor become radio-active. Each type of pottery emits its own characteristic gamma radiation enabling chemists to find out which chemical elements and how much of each were present in the pottery.

The application of these techniques to archaeology has become possible because of recently developed, highly sensitive silicon and germanium detectors and their associated equipment.

Application of Radiation to Processing

The application of ionizing radiation for the production or processing of chemicals and materials including the preservation of foods, the sterilization of biomedical supplies and in other areas is becoming widespread and important. While overall progress has been slow some significant applications have been made.

CALIFORNIUM-252 AS USED IN MINERAL EXPLORATION EXPERIMENTS

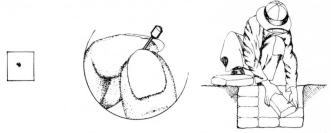

(1) Amount of californium used was about the size of the dot in the above box.

(2) Platinum foil compacted into a pellet about the size of a pencil eraser holds source.

(3) Sacks of low-grade ore from Nevada containing only 1/3 of an ounce of gold* per ton were stacked in shallow hole.

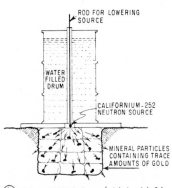

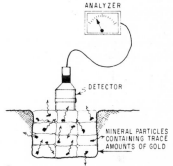

(4) Water filled drum (which shields personnel from neutron radiation) is placed over ore samples. Source rod is then pushed down to bring source in contact with samples. Neutrons emitted by californium atoms strike atoms of gold and are absorbed, making gold atoms radioactive.

(5) Radioactive gold atoms emit gamma rays, some of which strike detector which has been placed over ore samples after removal of californium source. Analyzer then determines amount of gold in ground minerals.

*Although gold is used as an illustration, silver was also detected by this technique, which is applicable to other minerals as well.

Figure 1-43. Use of californium-252 in mineral exploration.

Food preservation by irradiation

Many methods are available for preserving food for use during the non-harvesting times of the year and for use at large distances from the place of harvesting. Among these methods are: cooking, drying, smoking, salting, pickling, fermenting, canning, dry storage, freezing, refrigeration, sugar concentrates, and the use of chemical preservation. With the ready availability of relatively strong radiation sources, a new method of preserving food has become a real possibility.

Radiation preservation of food is accomplished either by *pasteurization,* by using low radiation doses of beta rays or gamma rays, or by *sterilization,* by using higher levels of these radiations. Radiation dosage is measured in units of the rad, the quantity of ionization which results in the absorption of 100 ergs of energy per gram of irradiated material. Pasteurization of food requires generally from 200,000 to 500,000 rads to be effective. The wide range is given because of the wide variety of foodstuffs involved and the results desired. At even lower doses, radiation can do some very effective things.

For example, a dose of 4,000 to 10,000 rads will inhibit the sprouting of potatoes or onions. *Figure 1.44* is a photograph showing six potatoes 16 months after being exposed to varying dosages of gamma rays. It will be noted in the photograph that the middle range of radiation best controls the sprouting. Grains and cereals can be disinfested of insects at irradiations of 20,000 to 50,000 rads. The larvae of insects that lodge within fruits can be sterilized by about 50,000 rads.

Food irradiation studies are being conducted using irradiation facilities at a number of sites in the United States as shown in *Figure 1.45*. In addition several irradiations have been loaned to other countries as shown in *Figure 1.46*. In general the studies are conducted so as to achieve optimum balance between the effective time the food can be preserved, the acceptability of the processed food as to colour, taste, odour and other characteristics, the general wholesomeness of the food for human consumption, and the overall economic benefits to be derived from the adaptation of radiation preservation.

Figure 1-44. Inhibition of sprouting of potatoes by radiation.

Seafood

Fish, like other foods from the sea, is a highly perishable product. Freshly caught shrimp have an ice-storage life of 14 to 16 days. If antibiotics or other chemicals are used, perhaps it can be kept fresh for up to 20 days. Crab meat normally has a 7 day life. Most of the fresh fish landed in Boston is sold within a 200 mile radius.

Studies have shown that shrimp, radiation pasteurized at 100,000 rads and maintained in crushed ice, were in apparent good condition after seven weeks. Under the same conditions unirradiated shrimp were spoiled after three weeks.

Flavour and odour of crab irradiated at 350,000 rads were still acceptable after about 40 days.

The Marine Products Development Irradiator has been built at Gloucester, Massachusetts to test the logistics and economics of

IRRADIATORS SUPPORTING THE FOOD-IRRADIATION EFFORT

Irradiator/Location	Description	Purpose
Research		
Massachusetts Inst. of Technology University of California (Davis) University of Washington University of Florida	Research 30,000-40,000 curies of ^{60}Co, pool type	On-site research
On-Ship No. 3 Seattle, Wash.	36,000 curies of ^{60}Co, 18-ton portable unit	Irradiation at sea
Portable Cesium No. 1 Various industries	170,000 curies of ^{137}Cs, 18-ton portable unit	Demonstration to and use by industry
Pilot Plant		
Marine Products Development Irradiator (MPDI), Gloucester, Mass.	250,000 curies of ^{60}Co, 1000 lb/hr at pasteurizing dose	Seafood irradiation
Mobile Gamma Irradiator (MGI), Davis, Calif.	100,000 curies of ^{60}Co, 1000 lb/hr at pasteurizing dose	Fruit irradiation
Grain Products Irradiator (GPI), Savannah, Ga.	25,000 curies of ^{60}Co, 5000 lb/hr bulk or 2800 lb/hr packaged	Grain products disinfestation
Hawaii Development Irradiator (HDI), Honolulu	225,000 curies of ^{60}Co, 4000 lb/hr at pasteurizing dose	Tropical fruits

Figure 1-45. Irradiators supporting the food irradiation effort.

IRRADIATORS ON LOAN TO OTHER COUNTRIES

Unit	Country	Date of Loan	Products
On-Ship No. 1	Israel	March, 1966	Citrus
	Chile	June, 1968	Potatoes, fish
On-Ship No. 2*	Iceland	April, 1968	Fish
Portable Cesium No. 2	Argentina	Spring, 1968	Bovine blood, fruit, fish
Portable Cesium No. 3	India	Spring, 1968	Wheat products
Research	Pakistan	Spring, 1968	Wheat products

* Formerly on the trawler Delaware based in Gloucester, Mass., USA.

Figure 1-46. Irradiators on loan to other countries.

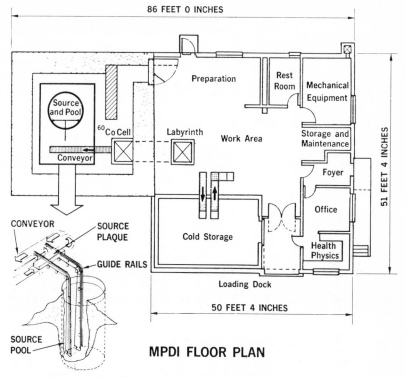

Figure 1-47. The marine products development irradiator.

radiation processing and to assist the fishing industry in com-
mercializing the process. This facility shown in *Figure 1.47*
houses a 250,000 curie[1] cobalt-60 source and can process 2000
pounds of seafood per hour with a 250,000 rad dose.

The first of the shipboard irradiators was installed aboard the
Delaware, a research fishing vessel operated by the U.S. Fish and
Wildlife Service of the Department of the Interior, shown in
Figure 1.48. Each of the irradiators shown in *Figure 1.49* and
Figure 1.50 weigh about 17 tons and use a 30,000 curie cobalt-60
source. These shipboard irradiators permit radiation preservation
studies to be carried out as soon after the fish are caught as possible.

---

1 A *curie* is the amount of radioactive material that produces 37,000,000,000
nuclear disintegrations per second.

Figure 1-48. The Delaware, a research fishing vessel.

Grain

A grain product irradiator is installed at the USDA's Entomological Research Station in Savannah, Georgia. This irradiator, utilizing an original 25,000 curie cobalt-60 source is capable of handling 5000 pounds of bulk grain per hour and is used for studies in the insect disinfection of grains and for test irradiation of packaged mixes, cereal and flour.

Fruit

The Hawaii Development Irradiator, located in Honolulu (see *Figure 1.51*) has a 250,000 curie cobalt-60 source and can process 4000 pounds of papaya, mango or other foods per hour at 75,000 rads.

Wholesomeness studies

Wholesomeness studies involve long term feeding of the specified irradiated food to selected experimental animals and detailed

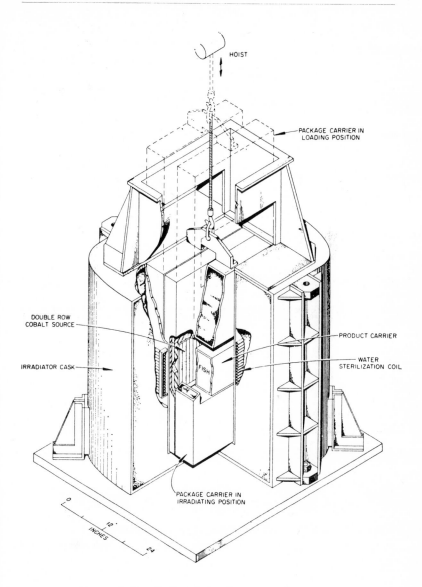

Figure 1-49. Shipboard irradiator, cutaway view.

Figure 1-50. Shipboard irradiator, photograph.

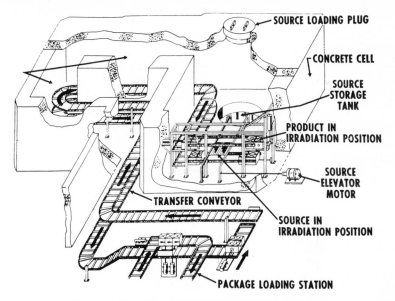

Figure 1-51. The Hawaii development irradiator.

examination of the results.   Before the U.S. Food and Drug Administration, the responsible agency, will approve any food for public consumption, they insist on specific evidence of wholesomeness, microbiological safety, absence of toxicity, nutritional adequacy, processing control, labelling and benefits.

The future of radiation preservations of foods in the U.S. is not yet certain.   Each food and each process is a separate case.   The ultimate use of irradiated food will depend on its general acceptance and its ability to compete with other methods of preservation. Before this can be done the public must have access to the produce.

Sterilization of biomedical supplies

The radiation emitted by cobalt-60, as well as that emitted by nuclear accelerators, can be used to sterilize surgical supplies and their plastic wrappers.   For example, sutures thus sterilized and wrapped are so much more convenient to the surgeon that radiation sterilized sutures have gained overwhelming acceptance.

Wood-Polymer Combination

Throughout the world a great deal of attention has been given to using radiation of wood-polymer combination to improve surface properties and esthetic appeal.

In these processes wood, impregnated with various organic monomer systems, like methyl methacrylate, is irradiated with relatively high dosages of gamma rays from a cobalt-60 source and cured in various ways. The resulting wood-polymer combinations are often superior in compression strength, hardness, water absorption, abrasion resistance and esthetic appeal to untreated woods.

In the United States three companies are currently using radiation to cure monomer-impregnated wood. The wood of one company has been specified for use as flooring for the Kansas City International Airport in which there is to be over ten acres of flooring.

Another company is in production and could produce more than a million square feet of a prefinished flooring in 1969. Technical and marketing development activities in wood-polymer materials have been under way at the third company for over three years. A fourth company is constructing a radiation facility designed for 5,000,000 curies of cobalt-60 and capable of processing 15-20,000,000 square feet of wood polymer annually.

Active and projected programmes in this field exist in at least 25 countries. Almost all of this work involves impregnating the fibrous substrate with a monomeric material and polymerizing it *in situ*. In many Far Eastern countries there is a desperate need for a cheap building board material which will stand up to the attack of insects, micro-organisms and severe weather conditions. There is hope that polymer-fibre board and similar combinations might help to fill the need.

Concrete polymer materials

For about three years one of the major efforts in the search for practical uses of radiation in industry has been the development and testing of concrete polymer materials. These materials are produced

Freeze - thaw Resistance

Control concrete, not impregnated. After 590 cycles in test, cylinder shows 26.5-percent weight loss.

Impregnated with methyl methacrylate. After 2520 cycles in test, cylinder shows 0.5-percent weight loss.

Figure 1-52. Freeze-thaw Resistance of concrete polymer.

by soaking ordinary hardened concrete, made of cement, water and sand and stone aggregate, in a liquid monomer such as methyl methacrylate. The material is then placed in a special chamber where it is exposed to a cobalt-60 radiation source for several hours. The liquid which has permeated the concrete, hardens into a plastic.

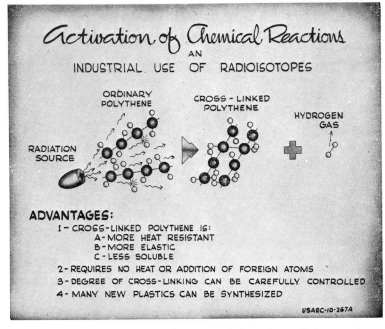

Figure 1-53. Crosslinking of polythene by radiation.

Ordinary hardened concrete contains about 14 percent of inter-connecting voids. When this structure is impregnated with about 6 percent by weight of monomer and the monomer is polymerized by exposure to radiation a three-dimensional polymer network within the existing concrete structure is produced. The compression and tensile strength of such concrete-polymer materials are increased markedly over the compression and tensile strengths of ordinary hardened concrete. Because the interconnecting voids are filled with polymer, water permeability is decreased to a negligible level. Consequently the new material exhibits drama-tically improved resistance to freeze-thaw damage, abrasion, and corrosion.

Figure 1.52 shows photographic evidence of the increased freeze-thaw resistance of a polymerized methyl methacrylate impregnated test cylinder as compared to control cylinders of ordinary concrete.

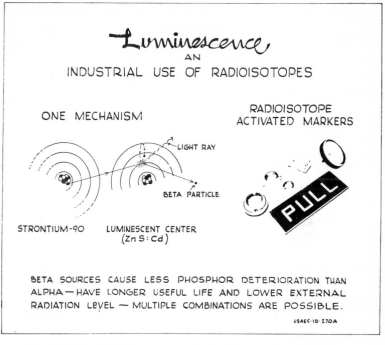

Figure 1-54. Luminescence, an industrial use of radioisotopes.

Crosslinking of Polyethylene

Radiation of polyethylene to produce crosslinking within the complex polyethylene molecule has produced special coatings for wire, insulating materials, heat shrinkable packaging film and bags of improved physical properties. Several companies report doing good business with these products. *Figure 1.53* illustrates the principle of crosslinking of polyethylene by radiation.

Grafting polymers via radiation

Starch graft copolymers are a new class of starch derivatives consisting of polyolefin chains grafted at frequent intervals to starch molecules. They are expected to have unique properties making them of potential application in the mining, textile, paper and petroleum industries.

These graft copolymers can be prepared in various ways but the use of high activity, high capacity cobalt-60 radiation is preferred because the free radicals generated by irradiation are located in the starch molecule itself. This seems to favour graft copolymerization.

Luminescence

Radiation from radioisotopes can cause some materials to become luminescent when they impinge on the materials. The glow from such luminescent materials is useful to mark visually important locations in the absence of illumination from other sources and when no suitable and reliable electrical source is available for lighting the marker. Radioisotope activated markers have wide uses in many situations, for example in the marking of emergency exits on ships and aeroplanes. *Figure 1.54* illustrates this use of radioisotopes.

CHAPTER 2

Nuclear Electric Power Development

The Fission Process

Otto Hahn and Fritz Strassman, working at the Kaiser Wilhelm Institute in Berlin, in the fall of 1938, were puzzled when they found the element barium in the residue material from an experiment in which they had bombarded uranium with neutrons.

They discussed their results with Lise Meitner who was then working with Niels Bohr in Denmark. Miss Meitner, and her nephew O. R. Frisch, suggested in January, 1939, that these experiments could be explained by saying that, "the uranium nucleus may after capture of a neutron divide itself into two nuclei of roughly equal size".

Within a few days of the publication of the views of Meitner and Frisch, Niels Bohr made a trip to the United States where he discussed the results with Albert Einstein and J. A. Wheeler, who had once been his student. The news spread rapidly for it was clearly indicative of the vast new developments to come.

Scientists throughout the world began an almost feverish activity to understand more about the phenomena. Within three months scientists were agreed that uranium nuclei could undergo fission when exposed to slow neutrons. The famous letter of Albert Einstein to President Roosevelt (*Figure 2.1*) brought the attention of government to the potentialities of this new process. The subsequent developments, leading first to the atomic bomb and later to the many peaceful uses, have been widely publicised in the past 25 years.

We now know that fissioning uranium nuclei split in many different ways yielding fission fragments of nuclei from those with atomic number $Z = 30$ (zinc), to $Z = 64$ (gadolinium), and having mass number from about $A = 72$ to about $A = 161$. *Figure 2.2* illustrates the process of fission. One of the many

Albert Einstein
Old Grove Rd.
Nassau Point
Peconic, Long Island

August 2nd, 1939

F.D. Roosevelt,
President of the United States,
White House
Washington, D.C.

Sir:

Some recent work by E.Fermi and L. Szilard, which has been communicated to me in manuscript, leads me to expect that the element uran
ium may be turned into a new and important source of energy in the immediate future. Certain aspects of the situation which has arisen seem
to call for watchfulness and, if necessary, quick action on the part
of the Administration. I believe therefore that it is my duty to bring
to your attention the following facts and recommendations:

In the course of the last four months it has been made probable -
through the work of Joliot in France as well as Fermi and Szilard in
America - that it may become possible to set up a nuclear chain reactio
in a large mass of uranium,by which vast amounts of power and large qua
ities of new radium-like elements would be generated. Now it appears
almost certain that this could be achieved in the immediate future.

This new phenomenon would also lead to the construction of bombs,
and it is conceivable - though much less certain - that extremely power
ful bombs of a new type may thus be constructed. A single bomb of this
type, carried by boat and exploded in a port, might very well destroy
the whole port together with some of the surrounding territory. However
such bombs might very well prove to be too heavy for transportation by
air.

-2-

The United States has only very poor ores of uranium in moderate quantities. There is some good ore in Canada and the former Czechoslovakia, while the most important source of uranium is Belgian Congo.

In view of this situation you may think it desirable to have some permanent contact maintained between the Administration and the group of physicists working on chain reactions in America. One possible way of achieving this might be for you to entrust with this task a person who has your confidence and who could perhaps serve in an inofficial capacity. His task might comprise the following:

a) to approach Government Departments, keep them informed of the further development, and put forward recommendations for Government action, giving particular attention to the problem of securing a supply of uranium ore for the United States;

b) to speed up the experimental work,which is at present being carried on within the limits of the budgets of University laboratories, by providing funds, if such funds be required, through his contacts with private persons who are willing to make contributions for this cause, and perhaps also by obtaining the co-operation of industrial laboratories which have the necessary equipment.

I understand that Germany has actually stopped the sale of uranium from the Czechoslovakian mines which she has taken over. That she should have taken such early action might perhaps be understood on the ground that the son of the German Under-Secretary of State, von Weizsäcker, is attached to the Kaiser-Wilhelm-Institut in Berlin where some of the American work on uranium is now being repeated.

Yours very truly,

A. Einstein

(Albert Einstein)

Figure 2-1. Letter Albert Einstein to President F. D. Roosevelt.

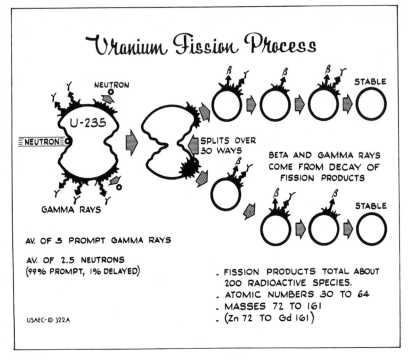

Figure 2-2. Uranium fission process.

equations which could be written to illustrate the fission process is given below:

$$_{92}U^{235} + {_0}n^1 \rightarrow {_{42}}Mo^{95} + {_{57}}La^{139} + 7 \text{ electrons} + 2 \text{ Neutrons} + \text{Energy} \tag{1}$$

Fission Products

The nuclei produced by fission almost invariably have too large a proportion of neutrons for stability. Each fragment thus initiates a short radioactive series involving the successive emission of negative beta particles. There are about 90 possible nuclei that can be formed by fission of an uranium nucleus and if most of these initiate chains of two or three successive members, the process of fission clearly must yield mixtures of very great complexities.

The yields of fission products from slow neutron fission of uranium-235, plotted against the mass number is shown in *Figure 2.3*. An examination of the fission yield curve shows that there are two peaks, at A = 95 and A = 139. Between these peaks at A = 117 the yield is a minimum and is about 600 times smaller than the yield at maximum.

About 97% of all U-235 nuclei undergoing thermal neutron fission yield product nuclei which fall into two groups, a *light* group with mass numbers from 85 to 104 and a *heavy* group with mass numbers 130 to 149. It is therefore clear that the thermal neutron fission of uranium-235 is far from symmetrical. One of the fragments is almost always somewhat larger than the other fragment.

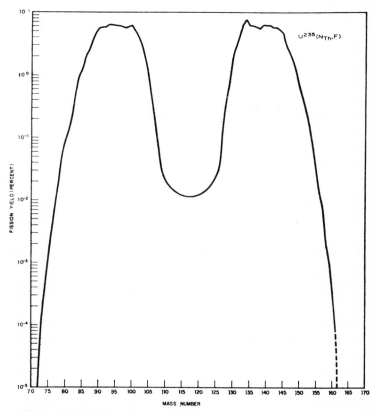

Figure 2-3. Yield of fission products from slow neutron fission of U-235.

The fission of a single uranium-235 nucleus in a reaction like that shown in equation (1) is accompanied by the release of more than 200 million electron volts* (MeV) of energy. The released energy is apportioned by the laws of conservation of momentum among the reaction products. *Table 2.1* lists the approximate distribution of fission energy to the various reaction products.

FISSION ENERGY DISTRIBUTION

Kinetic energy of fission fragment	165 Mev
Kinetic energy of neutron	5 Mev
Gamma rays emitted instantaneously	7 Mev
Energy released in subsequent radioactive decay	23 Mev
Total fission energy about	200 Mev

Table 2-1. Fission energy distribution.

It is evident from *Table 1* that the fissioning of a single uranium nucleus releases about $3 \cdot 2 \times 10^{-4}$ erg. The reader can verify that one watt of power is produced when 3×10^{10} uranium-235 nuclei undergo fission in one second. It can also be verified that the fission of one gram of uranium-235 corresponds to the production of 1,000 kilowatts of heat. To produce the same amount of heat from the combustion of coal would require burning more than 3 tons per day.

Fission Neutrons

A French team of scientists, F. Joliet, H. Von Halban, F. Perrin and L. Kowarski, established soon after the original discovery of fission that more than one neutron is emitted on the average by the fissioning of a nucleus of uranium-235. This was an important development because it stimulated scientists to believe that a divergent chain reaction might be possible.

Today we know that the number of neutrons emitted during a single fission reaction varies depending on the mode of fission. Sometimes only one neutron is emitted. In other cases two, three,

---

* An electron volt (eV) is the energy acquired by a charged particle carrying a unit electric charge when it falls through a potential of one volt. It is equivalent to $1 \cdot 6 \times 10^{-12}$ erg. Thus a million electron volts (MeV) is = to $1 \cdot 6 \times 10^{-6}$ erg.

four, or more may be released. The overall average is not an integer. *Table 2.2* records the average number of neutrons produced in the fissioning of three important isotopes.

Isotope					Average number of neutrons produced by slow neutron fission
U-233	....	....	....	....	2.50
U-235	....	....	....	....	2.43
Pu-239		....	....	....	2.89

Table 2-2. Neutrons per fission event.

The average number of neutrons released per fission increases with the energy of the neutrons inducing the fission. For example an average of 4·5 neutrons are produced per fission in U-235 by 14 Mev neutrons.

Most of the neutrons produced in fission are released in a very short time, in about 10^{-14} seconds, but a small number are not emitted for several seconds after the fission process has taken place.

The Chain Reaction

Since each fission reaction initiated by a single neutron striking a U-235 nucleus produces two or three neutrons of sufficient energy to cause other U-235 nuclei to undergo fission, a chain reacting system should be possible in principle. *Figure 2.4* illustrates the principle of fission and the chain reaction.

Even though on the average more than two neutrons are emitted per fission event, it is not a simple matter to fabricate a nuclear chain reactor. There are several ways that these neutrons can be lost from a real device.

First, some neutrons can simply leak out of the system and be lost. Secondly, some of the neutrons will be absorbed in the structural or other extraneous materials in the reactor. Thirdly, some of the neutrons may be captured in the uranium by a nuclear process which does not result in causing a uranium-235 nucleus to undergo fission. After all these losses of neutrons, it is necessary, if the chain reaction is to be maintained, that one neutron be left available for capture by another uranium-235 nucleus.

FISSION

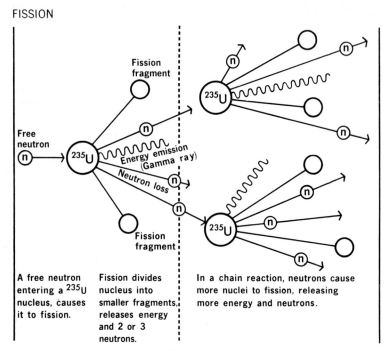

| A free neutron entering a ^{235}U nucleus, causes it to fission. | Fission divides nucleus into smaller fragments, releases energy and 2 or 3 neutrons. | In a chain reaction, neutrons cause more nuclei to fission, releasing more energy and neutrons. |

If a free neutron enters a ^{235}U nucleus, the nucleus will divide, or fission, into two smaller atoms, releasing energy and 2 or 3 more neutrons. These neutrons can then go on to cause more ^{235}U nuclei to fission.

Figure 2-4. Principle of fission and the chain reaction.

The First Reactor

Toward the end of 1942, a little more than three years after the announcement of the discovery of the fission process, sufficient amounts of pure uranium, graphite and other materials had been prepared to test the principle of the chain reaction. On the afternoon of December 2, 1942, a group of scientists at the University of Chicago, under the leadership of Enrico Fermi, an American of Italian origin, assembled these materials and for the first time in history a self-maintaining nuclear chain reaction was initiated. Photographs taken at several stages during the construction of the reactor are shown in *Figure 2.5, 2.6, 2.7* and *2.8.* An artist's sketch of the reactor is shown in *Figure 2.9.* By today's standards this

Figure 2-5. Construction of the world's first reactor (graphite layer forming the base).

Figure 2-6. Construction of the world's first reactor (seventh layer of graphite covering layer No. 6 containing 3" cylinder of black uranium oxide).

Figure 2-7. Construction of the world's first reactor (tenth layer of graphite and uranium.)

Figure 2-8. Construction of the world's first reactor (19th layer containing graphite).

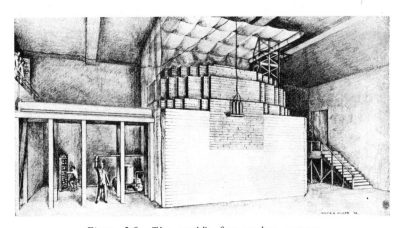

Figure 2-9. The world's first nuclear reactor.

Figure 2-10. The final resting place for the world's first nuclear reactor.

nuclear reactor was a crude apparatus — essentially an assembly of graphite bricks interspersed with chunks of uranium metal. The "pile", as it was called, was about 25 feet on the side and about 19 feet high. Primitive though it was, its successful operation ushered in a new era whose promise is yet unfolding.

The reactor was initially operated at about $\frac{1}{2}$ watt of total power. Later it was operated at 200 watts. In 1943 the first reactor was disassembled and rebuilt with certain refinements at the Argonne National Laboratory. In 1956 the graphite and uranium were removed and the remaining shell buried. A suitably inscribed monument now marks its resting place as shown in *Figure 2.10*.

Multiplication Factor

For each nucleus undergoing fission it is necessary, if a chain reaction is to be maintained, that there shall be produced at least one neutron which causes fission in another nucleus. It is convenient to express this rather complicated condition for the maintenance of a chain reaction by a single term called the "multiplication factor" of the system. This multiplication factor, k, is defined as the ratio of the number of neutrons produced by fission in any one generation to the number in the immediately preceding generation.

If k is exactly equal to unity for an assembly, a chain reaction is possible. In such a system if a particular fission generation starts with 1,000 neutrons in the system, there will be exactly 1,000 neutrons in the system at the beginning of each succeeding generation.

If k is greater than unity, the number of neutrons in the system will increase with time. If $k = 1 \cdot 005$, then 1,000 neutrons present in the system at the beginning of the first generation will increase to 1,005 neutrons at the beginning of the second generation. The number of neutrons would increase by another factor of $1 \cdot 005$ to 1,010 neutrons at the beginning of the third generation. And so the number of neutrons would grow and the number of fissioning nuclei would grow through all succeeding generations. This can result in uncontrollable growth unless some neutron absorber is introduced in the system to reduce the value of k to unity.

On the other hand, if k is less than unity the number of neutrons in the system will be progressively reduced in each generation and the chain reaction will not be propagated.

Critical Size

If the components of the multiplication factor are examined in some detail one finds that:

(a) the neutron leakage from the system is influenced in general by the size and shape of the system and in particular by the surface area of the system, and

(b) the neutron generation with the system is determined by the volume of the system and the composition and arrangement of its components.

To minimize the loss of neutrons by escape, it is therefore necessary to decrease the ratio of surface area to the volume of the system.

The value of the multiplication factor depends on the relative extent to which the neutrons take part in four processes:

(a) neutron leakage,

(b) non-fission capture by fuel nuclei,

(c) parasitic capture by various materials of construction, and impurities throughout systems including those produced by earlier nuclear reactions, (e.g. fission products by earlier nuclear reactions, (e.g. fission products accumulated in fuel), and

(d) fission capture of slow neutrons.

In all four of these processes neutrons are removed from the system.

Other neutrons must be generated by the fourth process to replace them if the chain reaction is to be sustained.

The critical size of a reactor is the size for which the number of neutrons lost by leakage and capture is just balanced by the number of neutrons generated by fission. The critical size is determined by many things, such as: isotopic composition of the uranium, its chemical impurities, the shape and arrangement of the fuel elements, the proportion and composition and location of the moderator coolants, the control rods, structural materials and the presence of various impurities and voids. Even the presence of a material around the reactor can act as a reflector of neutrons, returning them to the system rather than allowing them to escape.

If a system is smaller than the critical size, i.e., subcritical, neutrons are lost at a greater ratio than they are replenished by fission and so a self-sustaining chain reaction will be impossible.

Nuclear Cross Sections

In order to gain some quantitative measure of the probability of a nuclear reaction taking place, physicists have adopted the concept of a nuclear cross section. This represents the effective cross sectional area that a single nucleus of a given kind presents to a bombarding particle causing a particular reaction. The nuclear cross section for a certain reaction is a measure of the yield of that reaction.

If I is the number of neutrons striking, in a given time, a certain area of target material containing N nuclei per square centimeter, and A is the number of the nuclei that interact in a particular manner in the given time, then the neutron cross section σ for that process is defined by the expression:

$$\sigma = \frac{A}{NI} \text{ cm}^2/\text{nucleus} \qquad (2)$$

Experimental values for nuclear cross sections, while usually of the order of 10^{-24} cm^2/nucleus vary considerably depending on the probability of the reaction taking place. The fission cross section of uranium-235 for thermal neutrons is 577×10^{-24} square centimetre.

Calculation of Power of Fission Reactor

In a nuclear reactor at any instant there will be n neutrons per cubic centimetre moving with a velocity of v centimetres per second. The product nv, expressed in terms of the number of neutrons per square centimetre per second, is called the *neutron flux*. If there are N nuclei of uranium-235 per cubic centimetre in the reactor and if V is the volume of the reactor, then NV is equal to the total number of fuel nuclei in the reactor.

The total number of nuclei A undergoing fission per second in the reactor is given by

$$A = nv\,\sigma\,NV \qquad \text{fissions per second} \qquad (3)$$

We have seen earlier that $3 \cdot 1 \times 10^{10}$ fission per second produces one watt. Hence the power produced by a nuclear reactor is given by

$$P = \frac{nv\ \sigma\ NV}{3\cdot 1 \times 10^{10}} \quad \text{watts} \qquad (4)$$

It is often more convenient to express this in terms of the mass of fuel in the reactor. This can be done by recalling that the mass, m, of the fuel is given by:

$$m = \frac{235\ NV}{6\cdot 02 \times 10^{23}} \quad \text{grams} \qquad (5)$$

and that the cross section for fission of uranium-235 by thermal neutron is 577×10^{-24} cm^2.

Inserting these values in equation (4), it is found that:

$$P = 4\cdot 8 \times 10^{-11}\ m\ (nv) \qquad \text{watts} \qquad (6)$$

It can be seen from (6) that the power output of a given reactor, containing m grams of fissionable material is proportional to the neutron flux (nv).

Reactor Uses

Since that day in December 1942 when man first constructed an atomic reactor, thousands of chain reacting systems have been constructed throughout the world. Many uses have been found for them and much has been learned about how to build reactors economically and how to operate them successfully.

In the first few years after reactors were invented, the attention of scientists and engineers was directed toward finding various uses for these devices. Almost from the start it was clear that reactors would be useful in research laboratories as intense sources of nuclear radiation for experimental purposes. *Figure 2.11* shows the control room of one such reactor, the Brookhaven National Laboratory High Flux Beam Research Reactor (HFBR). This reactor has a maximum neutron flux of $1\cdot 6 \times 10^{15}$ neutrons per square centimetre per second and a maximum thermal power of 40 megawatts. *Figure 2.12* is a photograph of the Experimental Floor of this reactor where nine beam holes allow the neutrons to escape through the eight foot thick concrete biological shield to various experimental systems. Several neutron spectrometers and choppers, together with auxiliary equipment are seen installed at the beam holes.

Figure 2-11. High flux beam research reactor control room (HFBR).

Figure 2-12. High flux beam research reactor experimental floor (HFBR).

Figure 2-13. Removal of radioisotope from Brookhaven graphite research reactor.

Reactors of this type are often used to produce radioisotopes for uses described in more detail in another chapter in this series. *Figure 2.13* shows a small radioactive sample being removed from one of the pneumatic tubes at the Brookhaven Graphite Research Reactor. With this facility small amounts of material are quickly blown in to the reactor by air pressure. After the required time in the reactor the sample is returned by reversing the air pressure. The health physicist at the right is checking the radiation level with a portable meter.

Table 2.3 lists some of the areas of science which are being explored by scientists using fission reactors of one kind or another. Some of the subjects being studied are also listed. *Figure 2.14* shows the rapid growth in the number of nuclear reactors in the United States for the period 1955-1968. Especially noteworthy is the large number of research and training reactors and critical experimental facilities. This illustrates quite vividly the rate and scale of nuclear research in the United States.

REACTORS SERVING SCIENCE

Area of Science	Subject
Nuclear physics	Study of nuclear reactions by irradiation target materials.
Solid state physics	Determination of crystal structures of materials by neutron diffraction techniques.
Radiation chemistry	Study in the effect of radiation on chemical reactions and on the properties of material.
Analytical chemistry	Measuring trace amounts of impurities in materials by activation analysis.
Agriculture	Development of new crops by inducing of genetic mutations by irradiation techniques.
Biology	Use of radioisotopes in studying the process of life.
Medicine	Use of radiation in therapy and diagnosis.
General	Production of radioisotopes.
Engineering	Testing of materials and processes.
Industry	Development of electrical power.

Table 2-3. Reactors serving science.

Nuclear Reactor Technology

Nuclear reactors based on the fission chain reaction generally consist of five basic components: Fuel System, Moderator, Control System, Energy Extraction System and Biological Shield. Let us examine each of these components to understand better the impact of their properties on reactor design parameters.

Fuel System

For a nuclear reactor to operate it must have in its fuel system an adequate amount of fissionable material — that is, some material that undergoes fission when struck by a neutron. The three known fissionable materials are uranium-235, plutonium-239 and uranium-233.

Fuel materials

Uranium is the only natural substance which contains sufficient amounts of fissionable materials to be of practical importance. The fissionable component of natural uranium is the isotope — uranium-235. Only about $0 \cdot 71\%$ of natural uranium is uranium-235. Almost all the rest is uranium-238. At first glance this small amount of uranium-235 may not appear significant. But it

GROWTH OF LICENSED NUCLEAR REACTORS AND FACILITIES

DECEMBER 31, 1955 - DECEMBER 31, 1968

Licenses in Effect for Operation or Possession of Reactors and Other Facilities

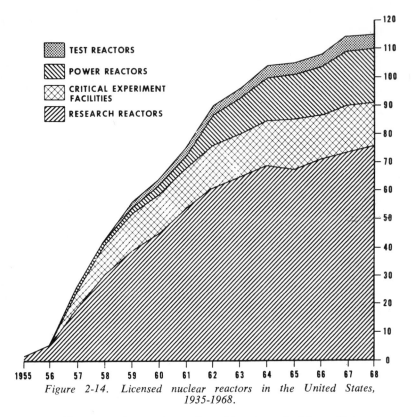

Figure 2-14. *Licensed nuclear reactors in the United States, 1935-1968.*

was with natural uranium that Fermi and his collaborators demonstrated in 1942 that a chain reacting system could be constructed.

Although uranium-238 does not undergo fission when struck by a slow neutron, another important reaction does occur. The U-238 nucleus absorbs a neutron and is transformed into uranium-239, a radioactive isotope of uranium. The reaction is:

$$_{92}U^{238} + {}_0N^1 \longrightarrow {}_{92}U^{239} + \text{gamma ray} \qquad (2$$

E. M. McMillan and P. H. Abelson, in 1940, showed that the U-239 nucleus decays spontaneously, with a $2 \cdot 3$ minute half-life. The U-239 nucleus decays by ejecting a beta particle and changing into an isotope of a new element which they called neptunium for the planet, Neptune, the next one out from Uranus.

The neputunium nucleus is also unstable and decays with a $2 \cdot 3$ day half-life, by emitting a second beta particle. In doing this it transforms itself into yet another new element, called plutonium by its discoverers because it was the second element beyond uranium as the planet Pluto is the second planet beyond Uranus.

The reactions are:

$$_{92}U^{239} \xrightarrow[\text{2·3 min.}]{\text{decay}} {}_{93}\text{Neptunium}^{239} + e^- \qquad (3)$$

and

$$_{93}\text{Neptunium}^{239} \xrightarrow[\text{2·3 day}]{\text{decay}} {}_{94}\text{Plutonium}^{239} + e^- \qquad (4)$$

Plutonium 239 decays rather slowly — its half life is 24,400 years — by the emission of an alpha particle. The decay reaction is

$$_{94}\text{Plutonium}^{239} \xrightarrow[\text{24,400 years}]{\text{decay}} {}_{92}U^{235} + {}_{2}He^4 \qquad (5)$$

The discoverers of plutonium-239 found it to be capable of undergoing the fission process when struck by a slow neutron. Thus by making use of reactions (2), (3) and (4), we can produce a new fissionable material in the course of reactor operation using natural uranium as a fuel. In such a reactor nuclei of the fission-able material uranium-235 are consumed, energy (either useful or waste) is released to the environment, and, at the same time, nuclei of plutonium-239 are produced. Since plutonium is different chemically from uranium, it is relatively easy to separate the plutonium from the uranium fuel elements.

Large industrial plants have been constructed in the United States, United Kingdom, France, and the Soviet Union for the production of large quantities of plutonium. *Figure 2.15* shows the front face of a Hanford, Washington production reactor. Operators are shown getting ready to "charge" fuel elements of natural uranium into one of the more than 3,000 process tubes. As fresh fuel elements are inserted, the irradiated fuel elements

Figure 2-15. Front face of a Hanford production reactor.

are ejected from the discharge end of the tube at the rear face. *Figure 2.16* is a photograph of one of the plutonium production reactors at Savannah River, Georgia.

Another fissionable material, uranium-233, can be produced by similar reactions by the neutron bombardment of the element thorium.

It is possible, by appropriate reactor design and operating techniques, to achieve a net gain of fissionable materials in a nuclear reactor. These are called "Breeder" reactors because, over a long period of time they can make more fissionable material than the amount consumed during the process. Because of their importance a special treatment is given in another part of this chapter.

Fuel Elements

The amount of fissionable material in the fuel used in a nuclear reactor has an important bearing on the physical size of the reactor. The reactor designer will specify a high concentration

Figure 2-16. One of the production reactors at Savannah River, Georgia.

of fissionable material in the fuel if he wants a compact reactor for, say, application to a space vehicle. Where compactness is not necessary or desirable, a less concentrated fuel may be used.

Fissionable material may be used in many forms as the fuel for a nuclear reactor. Although in recent years most reactor operating experience has been with solid fuel it is possible to operate reactors with fluid fuels. For example an aqueous solution of enriched uranium has been used as a fuel for a reactor.

Solid fuel elements may be made from metallic uranium or plutonium or from one or more of their solid compounds, or from numerous mixtures and alloys with other materials. Depending on the specific design objectives the solid fuel elements may be fabricated into cylindrical rods, spheres, plates, pins and complex or simple fuel assemblies. Often the fuel elements are enclosed in a protective coating to reduce the corrosive action of the reactor coolant and to contain the fission products within the fuel element rather than allow them to contaminate the other reactor parts. Stainless steel, aluminium and zirconium alloys are

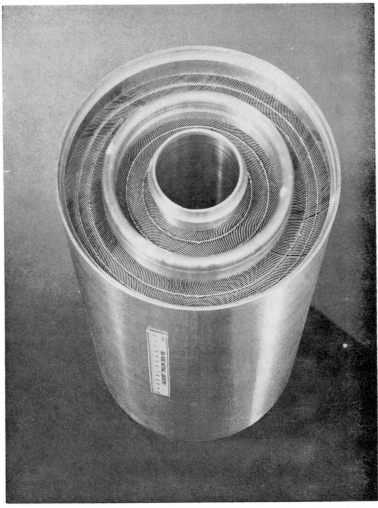

Figure 2-17. HFIR core with the two elements nested as in the reactor.

commonly used as cladding materials. Some specific fuel elements are discussed below so that the wide variety of design can be appreciated.

Figure 2.17 is a photograph of the fuel element of the High Flux Isotope Reactor (HFIR) at the Oak Ridge National

Laboratory. The active fuel material is contained in the thin metal plates. A coolant flows between these plates to remove the heat generated by the chain reaction. Note the necessity for careful and detailed design and fabrication techniques. *Figure 2.18* is a cutaway view of the central core of the HFIR showing the placement of the fuel element in the assembly. The reactor is specially designed to provide a high flux of neutrons in the central cylindrical core of the system. The high flux, about 5×10^{15} neutrons/cm²/sec, is used for the irradiation and preparation of special samples needed for studying the properties of the trans-plutonium elements.

Figure 2.19 is a photograph of a typical modern pressurized water reactor fuel assembly. Note the size and careful design of such a fuel element. This type of fuel element is currently used in the commercial generation of electric power and therefore must stand up during long periods of use at high temperature, and in extremely high radiation environments. Unless fuel elements are properly made they can fail.

Figure 2.20 illustrates a type of fuel element failure. Here the fuel element shown is one from the BONUS power reactor experiment. A longitudinal rupture of the sheath surrounding the active fuel material has made the fuel element useless.

Figure 2.21 shows a photograph of another fuel specimen illustrating the failure of the cladding after the specimen had been irradiated in a reactor for a long time.

The failure of a fuel element is not always as obvious as those failures shown in *Figure 2.20* and *2.21*. *Figure 2.22* shows a photograph of the cross section of a fuel element used in the EBR-2 experiment. This photograph illustrates the internal failure of the cladding material. An external examination of this fuel element might not reveal the true condition of the fuel material within. Such conditions as are shown by microscopic examination of this sample can result in serious failure.

As one can observe from *Figure 2.17* and *2.19* fuel elements must be designed carefully and fabricated to close dimensional tolerance. Often a few thousandth of an inch variance in a critical dimension can result in a fuel element becoming stuck in its channel or the flow of coolant can be retarded or blocked. Serious and

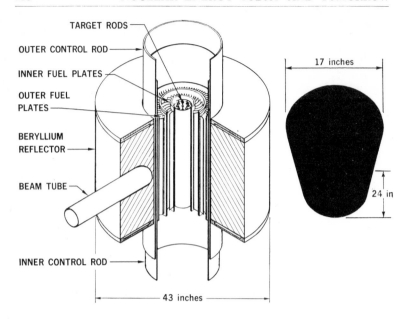

TARGET RODS

OUTER CONTROL ROD

INNER FUEL PLATES

OUTER FUEL PLATES

BERYLLIUM REFLECTOR

BEAM TUBE

INNER CONTROL ROD

17 inches

24 in

43 inches

CENTRAL CORE OF THE HIGH FLUX ISOTOPE REACTOR

Figure 2-18. Central core of the HFIR.

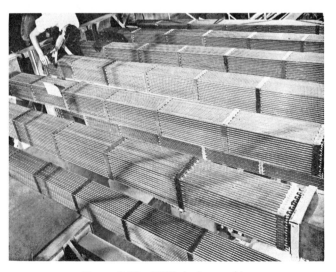

Figure 2-19. PWR fuel assembly.

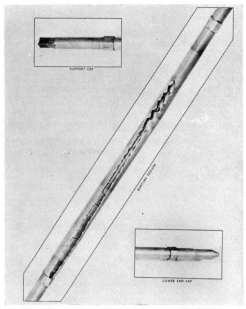

Figure 2-20. A failed fuel element (from the BONUS reactor).

Figure 2-21. Fuel specimen illustrates cladding failure.

costly operating and safety problems can then be encountered in the continued operation of the reactor.

Obviously fuel element fabrication techniques must be adopted that prevent, insofar as practical, the failure of fuel elements under reactor operating conditions.

The fuel element is the very heart of a nuclear reactor and its

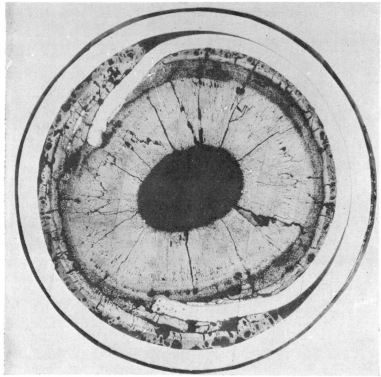

Figure 2-22. Internal failure of cladding material magnification.

design and performance determine to a large extent the overall economic competitiveness of nuclear reactors.

Considerable fuel element development work is required before any one particular fuel element assembly can be adapted for a reactor. Prior to this development work there must be a firm foundation laid for understanding the basic properties of the materials to be used in reactor construction. Taken together the research and development work on fuel development represents a major activity in nuclear reactor studies.

A current problem of great concern is the formation of tiny voids within materials like stainless steel, aluminum and nickel when they are radiated for long periods at moderately high temperature and at high neutron fluence. These voids tend to grow in size and

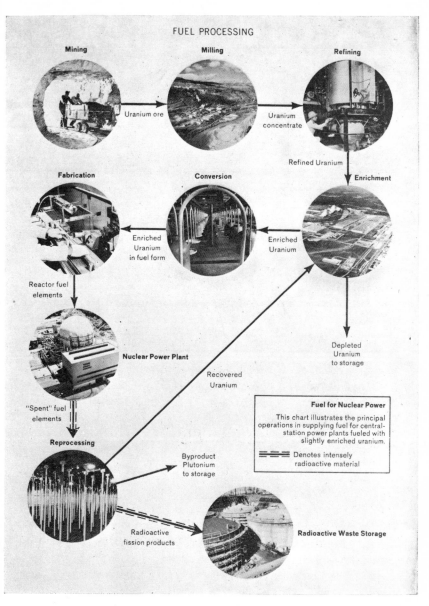

Figure 2-23. Steps in the supply of atomic fuel.

produce dimensional changes of significant magnitude in materials from which the reactors are constructed. Much effort to understand the mechanism involved and how to overcome the effects is underway.

Fuel cycle

The principal operations involved in preparing and supplying fission fuel for use in reactors are illustrated in *Figure 2.23*. In general after the uranium ore is removed from its deposit in nature it is milled and refined. Purified uranium compounds are processed through an isotopic enrichment plant where the uranium-235 concentration is raised from its normal value of $0 \cdot 71\%$ to the percentage enrichment required for the particular reactor in which it is to be used. Carefully designed and fabricated fuel elements are installed in the reactor and the reactor operated until the fuel elements are no longer useable. Upon removal from the reactor the fuel elements are reprocessed chemically to remove the radioactive fission products. Provisions are made for the proper and safe disposal of radioactive waste products. The recovered uranium is then returned to the fabrication plant to be re-used in other fuel elements. The only step in the nuclear fuel cycle that is not in the hands of private industry in the United States now is the operation of the isotopic enrichment plant now done entirely by the government. However, the nuclear industry can now have its own uranium enriched in the government plants by paying appropriate toll charges.

Moderator

The neutrons produced in a fission reaction like that shown in equation (1), are quite energetic and many have energies in the millions of electron volts range. They lose energy as they collide with surrounding matter in the reactor core. This loss of energy is desirable because slow moving neutrons are more effective than are fast neutrons in causing uranium-235 nuclei to undergo fission. Most reactors are maintained by fissions produced in the fuel nuclei when they absorb slow neutrons. It is often necessary that some material, called a moderator, be distributed throughout the reactor, especially within the fuel zone, to slow down the high energy neutrons by repeated elastic scattering collisions.

Table 2.4 shows the average number of collisions which a 2 MEV neutron must undergo with several typical nuclei before the energy of the neutron is reduced to about 0·025 electron volts (about the average value for a slow neutron). It is thus evident why the light elements like hydrogen, carbon and beryllium make good moderators because their nuclei absorb more energy from colliding with neutrons than do the nuclei of the heavier elements.

SCATTERING PROPERTIES OF NUCLEI

Nuclei			Mass No.	Average number of collisions to reduce neutron energy from 2 Mev to 0.025 ev
Hydrogen	....	....	 1	18
Deuterium	....	....	 2	25
Helium	....	....	 4	43
Beryllium	....	....	 9	86
Carbon	....	....	 12	114
Uranium	....	....	 238	2,172

Table 2-4. Scattering properties of nuclei.

Control system

To keep the number of fissioning nuclei within manageable limits in a reactor it is necessary to regulate the number of neutrons that swarm about in the core of the reactor. One can do this by several ways.

Perhaps the most widely used method of reactor control is the insertion of a material into the reactor which has the property of absorbing neutrons. Rods of such material are usually equipped with mechanisms which can regulate the depth of the insertion and the speed of movement so that control can be maintained. A second set of rods, called safety rods, are usually provided to permit rapid shutdown of the reactor in an emergency.

Energy extraction system

The heat produced in the fuel elements of a reactor can be removed by well known heat transfer techniques. If the amount of heat is trivial, it may be allowed to dissipate itself. If the amount of heat is moderate, a coolant system based on natural convection of a coolant like ordinary water or air may be sufficient.

Figure 2-24. World's first use of nuclear electric power —
December 20, 1951.

In nuclear electric power production, and other large scale applications, the heat is the primary product of the reactor. This heat must be conveyed out of the reactor through elaborate heat extraction mechanisms. Materials such as, liquid metals, organic liquids, fused salts, high pressure water, etc. are used to remove the heat from the reactor core.

Biological shield

Nuclear reactors contain large amounts of penetrating nuclear radiation and therefore must be heavily shielded to protect personnel from radiation exposure. This shield usually takes the form of several feet of concrete surrounding the reactor installation.

Nuclear Electric Power — Early Developments

In the first five or six years after the end of World War II emphasis in the United States was placed on determining the feasibility of generating electricity from the atom and on developing a basic background in science and technology. There were many enthusiasts and many sceptics as to the role which the atom might play in the development of electrical energy from fission reactors.

Feasibility of generating electrical power

The production of a significant amount of electric power on December 20, 1951, by the Experimental Breeder Reactor No. 1 (EBR-1), a sodium-potassium alloy cooled, fast reactor, demonstrated that it was possible to use the chain reaction to generate electrical power. The core of EBR-1 consisted of a number of fuel rods of highly enriched U-235 clad in stainless steel, a blanket of natural uranium surrounded the core and control was achieved by moving sections of the blanket. *Figure 2.24* is a photo of the first known use of electric power from atomic energy. The lamps are lighted from electricity produced by the generator on the right which received its heat from the Experimental Breeder Reactor. After operating for some time, a later careful analysis of the EBR-1 results showed that the efficiency for regeneration of fissionable material was slightly over 100 percent and that the principle of breeding was valid. The EBR-1 continued to operate until 1956 when the nuclear core was destroyed by a power excursion and operation was stopped until late 1957. The experiment was completed and the reactor shut down in 1964.

Nuclear propulsion of naval vehicles

The development of nuclear systems for the propulsion of submarines in the U.S.A., and for the propulsion of an ice breaker in the Soviet Union, in the early 1950's made it clear to any remaining sceptics that the atom would have a role to play in providing man with a new source of energy. *Figure 2.25* shows the nuclear powered submarine SKATE breaking through the ice during a trip under the Polar ice cap. Such endurance feats are impossible with conventional fuelled submarines.

International developments

On December 8, 1953, President Eisenhower offered a proposal to the United Nations General Assembly for developing peaceful uses of atomic energy. On that occasion the President said:

Figure 2-25. The nuclear submarine "Skate" breaks through the Polar ice cap.

"I would be prepared to submit to the Congress of the United States, and with every expectation of approval, any . . . plan (for international co-operation) that would:

"encourage world - wide investigation into the most effective peacetime uses of fissionable material, and with the certainty that they had all the material needed for the conduct of all experiments that were appropriate;

"begin to diminish the potential destructive power of the world's atomic stockpiles;

"allow all peoples of all nations to see that, in this enlightened age, the great powers of the earth, both of the East and of the West, are interested in human aspirations first, rather than in building up the armaments of war;

"open up a new channel for peaceful discussion, and initiate at least a new approach to the many difficult problems that must be solved in both private and public conversations, if the world is to shake off the inertia imposed by fear, and is to make positive progress toward peace.

"To the making of these fateful decisions, the United States pledges to devote its entire heart and mind to find the way by which the miraculous inventiveness of man shall not be dedicated to his death, but consecrated to his life."

This offer and the response of the Soviet Union and other great nations throughout the world gave nuclear power great impetus and developments were rapid. Discussions were begun that culminated in 1957 in establishing the International Atomic Energy Agency (IAEA). A new law was adopted in 1954 in the United States permitting private industry to own their reactors and thus to participate more effectively in Atomic Energy Developments. The same law permitted the government to exchange certain technical data with other countries and to adopt provisions which in general would promote the development of a private atomic energy industry in the United States. Similar co-operative steps were taken in most other countries of the world including the Soviet Union.

An Early Plan for Development of Nuclear Power

Soon after the President's call for peaceful atomic development, five nuclear power reactor projects were initiated.

Shippingport

The Pressurized Water Reactor, a reactor of a type originally designed for the nuclear propulsion of submarines, was built at Shippingport, Pennsylvania. This reactor employed fuel elements which had to be developed from a new alloy of zirconium and aluminium, and the fabrication of the world's largest pressure vessel up to that time. The principle of this reactor was to employ a pressurized water system to remove the heat from the reactor at temperatures high enough for efficient steam generation in a second supply. The steam in the second loop turns a turbine generator to produce electricity. See *Figure 2.26* for a schematic representation of a pressurized water system.

The Shippingport Station was a joint project of the U.S. Atomic Energy Commission, the Duquesne Light Company (an operating utility company) and Westinghouse Electric Corporation. Westinghouse under contract to the AEC designed and developed the nuclear reactor and Duquesne Light built the electric generating portion of the plant and served as operator for the entire Station.

Construction on this unit started in April 1955. It was completed

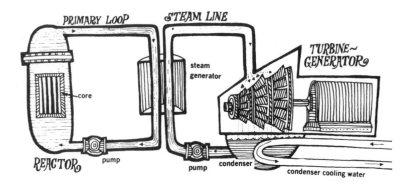

Figure 2-26. Schematic representation of a pressurized water system.

Figure 2-27. Lowering the core of the Shippingport atomic power station into position.

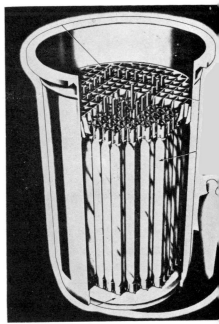

Figure 2-28. A cutaway view of the Shippingport reactor core.

*Figure   2-29.   World's   first   full-scale   atomic   power   plant —
Shippingport,   Pennsylvania.*

on schedule and was delivering power before the end of 1957.
Originally designed for producing 60,000 kwe it was modified and
resumed operation in 1965 at 90,000 kwe. *Figure 2.27* is a
photograph showing engineers lowering the core of the Shippingport
Reactor into position.   The fuel charge at this time consisted of 14
tons of natural uranium and 165 pounds of highly enriched uranium.
Figure 2.28 is a diagram showing the interior of the core.   Fuel
assemblies are locked into the top grid and bottom plate.   The close
array forms a critical mass.   The resulting chain reaction generates
heat which is removed by water flowing upward through channels.
Control rods are shown at the top of the picture. *Figure 2.29* is a
photograph of the Shippingport Power Station the world's first full-
scale atomic-electric power plant devoted exclusively to serving
civilian needs.

Sodium reactor experiment (SRE)

The Sodium Reactor Experiment was designed to take advantage of extensive experience with graphite reactors and of excellent heat transfer properties of a liquid sodium coolant; considerable experience was gained from this experiment.

Experimental boiling water reactor (EBWR)

The use of high pressure equipment required in pressurized-water reactors like the one built at Shippingport is costly. It is well known that if boiling can be permitted, water is much more effective for heat removal than if there is no boiling. Research at the Argonne National Laboratory also indicated that the stable operation of a power reactor might be achieved even if the water coolant were allowed to boil in the reactor. *Figure 2.30* illustrates the principle of a boiling water system for nuclear reactors. To test these concepts plans were made to construct a 5,000 kwe nuclear reactor at the National Reactor Testing Station in Idaho, to be completed at the end of 1956. Following the completion of this experiment on schedule, a series of boiling reactor experiments began in 1957. The experiments were terminated in 1967.

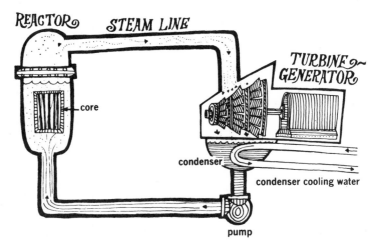

Figure 2-30. Boiling water system.

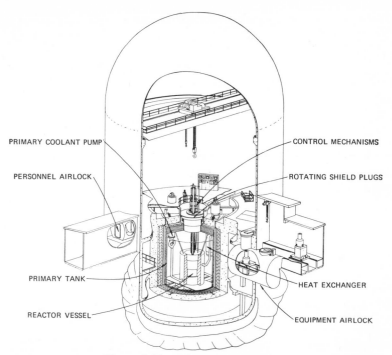

Figure 2-31. Schematic of EBR-2.

Figure 2-32. Reactor core of EBR-2.

Experimental breeder reactor No. 2 (EBR-2)

The second Experimental Breeder Reactor (EBR-2) was designed to demonstrate efficient breeding, high thermal efficiency, and the use of prototype components suitable for central station power plants. To achieve these aims liquid sodium was chosen as the coolant. The reactor was designed to produce 15,000 kwe of electricity and the plant included a complete fuel processing and fuel fabrication facility.

Figure 2.31 is a schematic of the Experimental Breeder Reactor No. 2. *Figure 2.32* is a view of the core of the reactor.

The reactor core and the entire primary system are submerged in sodium contained in a tank 26 feet in diameter and 26 feet deep. At full power the sodium coolant inlet temperature is 700°F and the outlet temperature is 883°F.

This reactor has been of great value in giving the research engineer a broad general understanding of the behaviour of such systems and specific detailed knowledge which will assist him in designing future full scale plants.

Homogeneous reactor experiment No. 2

A chain reacting system can be constructed using a slurry, or a solution, of the fissionable material in water. At the Oak Ridge National Laboratory a considerable amount of experience has been obtained with such systems. The Homogeneous Reactor Experiment No. 2, a greatly improved version of an earlier reactor experiment, was scheduled to operate at thermal power levels of about 10,000 kilowatts. Construction of the reactor was completed on schedule but a series of technical difficulties especially leaks, caused by stress corrosion of the vessel walls by the active fluid delayed power operation until 1957. When the tests were completed, scientists learned enough from this experiment to direct their attention elsewhere. In particular, this experiment was an excellent demonstration of the need for much basic information on the nature and causes of material failure under the extreme environmental conditions found in operating a reactor.

Industrial Development of Nuclear Power

Concurrent with the technological development by the U. S. Government of these five reactor units, American industry began its participation in the commercial development of electrical power from the fission process. At first the costs of development were shared by government and by industry but, as the reactors became more and more commercially attractive, industry has taken more and more of the full responsibility for the programme.

In general two types of reactors have been chosen by industry as being competitive, at the time the choice was made, with conventional fuelled power plants. These are the pressurized water reactors and the boiling water reactors.

Dresden nuclear power station

The Commonwealth Edison Company, a public utility serving the Chicago area, built a boiling water reactor (BWR) with 200,000 kwe capacity and began commercial operation in 1960.

Figure 2-33. Dresden Nuclear Power Station, Unit 1.

This unit, located at Morris, Illinois, about 50 miles southwest of Chicago, was designed by the General Electric Company for the owner-operator, the Commonwealth Edison Company, an operating utility company.

A photograph of this nuclear unit which is still in operation is shown in *Figure 2.33*.

Yankee atomic electric power station

The Yankee Electric Power Company and the Westinghouse Electric Corporation designed, developed and constructed a pressurized water reactor (PWR) of 175,000 kwe capacity at Rowe, Massachusetts about 45 miles east of Albany, New York in 1960. A photograph of this, the third large nuclear power plant to be built in the United States, is shown in *Figure 2.34*.

Status of Nuclear Power Plants in the United States

After going through periods of enthusiasm and periods of disappointment, the utility companies and the large nuclear reactor

Figure 2-34. The Yankee Atomic Electric Power Station.

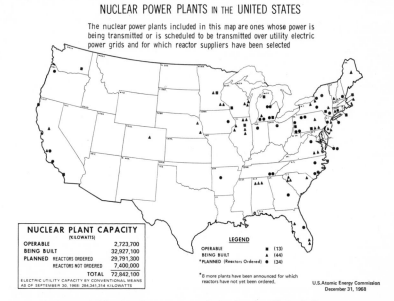

Figure 2-35. Nuclear power plants in the United States.

design organizations of the United States seem to have reached a conclusion that new nuclear power plants of large capacity are sometimes competitive with conventional coal fired power plants. Accordingly, new construction of nuclear electrical power plants is growing rapidly.

Figure 2.35 shows the current status of nuclear power plants in the United States. Only those plants are shown on the map whose electrical power is being transmitted over utility electric grids, or, those plants scheduled for transmitting electrical power on the grid and for which the nuclear reactor supplier has been selected. At the time the statistics were prepared, January 1, 1969, there were (a) 13 nuclear power plants operable with a total capacity of 2,723,700 electrical kilowatts, (b) 44 plants with total generator capacity of 32,927,100 electrical kilowatts being built, and (c) 42 plants planned with total generator capacity of 37,291,300 electrical kilowatts.

	SITE	CAPACITY (Kilowatts)	UTILITY	STARTUP
ALABAMA	Browns Ferry	1,064,500	Tennessee Valley Authority	1970
	Browns Ferry	1,064,500	Tennessee Valley Authority	1971
	Browns Ferry	1,064,500	Tennessee Valley Authority	1972
ARKANSAS	Dardanelle Lake	850,000	Arkansas Power & Light Co.	1972
CALIFORNIA	Humboldt Bay	68,500	Pacific Gas & Electric Co.	1963
	San Clemente	430,000	Southern Calif. Edison and San Diego Gas & Electric Co.	1967
	Corral Canyon	462,000	L. A. Dept. of Water & Power	1974
	Diablo Canyon No.1	1,060,000	Pacific Gas & Electric Co.	1972
	Diablo Canyon No. 2	1,070,000	Pacific Gas & Electric Co.	1974
	Sacramento County	800,000	Sacramento Municipal District	1972
COLORADO	Platteville	330,000	Public Service Co. of Colorado	1971
CONNECTICUT	Haddam Neck	462,000	Conn. Yankee Atomic Power Co.	1967
	Waterford No. 1	652,100	Northeast Utilities	1969
	Waterford No. 2	828,000	Northeast Utilities	1973
FLORIDA	Turkey Point No. 3	651,500	Florida Power & Light Co.	1970
	Turkey Point No. 4	651,500	Florida Power & Light Co.	1971
	Red Level	825,000	Florida Power Corp.	1972
	Hutchinson Island	800,000	Florida Power and Light Co.	1973
GEORGIA	Baxley	786,000	Georgia Power Co.	1973
ILLINOIS	Morris No. 1	200,000	Commonwealth Edison Co.	1959
	Morris No. 2	715,000	Commonwealth Edison Co.	1968
	Morris No. 3	715,000	Commonwealth Edison Co.	1969
	Zion No. 1	1,050,000	Commonwealth Edison Co.	1972
	Zion No. 2	1,050,000	Commonwealth Edison Co.	1973
	Quad Cities No. 1	715,000	Comm. Ed. Co.—Ia.—Ill. Gas & Elec. Co.	1970
	Quad Cities No. 2	715,000	Comm. Ed. Co.—Ia.—Ill. Gas & Elec. Co.	1971
INDIANA	Burns Harbor	515,000	Northern Indiana Public Service Co.	1970's
IOWA	Cedar Rapids	545,000	Iowa Electric Light and Power Co.	1973
MAINE	Wiscasset	790,000	Maine Yankee Atomic Power Co.	1972
MARYLAND	Lusby	800,000	Baltimore Gas and Electric Co.	1973
	Lusby	800,000	Baltimore Gas and Electric Co.	1974
MASSACHUSETTS	Rowe	175,000	Yankee Atomic Electric Co.	1960
	Plymouth	625,000	Boston Edison Co.	1971
MICHIGAN	Big Rock Point	70,300	Consumers Power Co.	1962
	South Haven	700,000	Consumers Power Co.	1969
	Lagoona Beach	60,900	Detroit Edison Co.	1963
	Lagoona Beach	1,100,000	Detroit Edison Co.	1974
	Bridgman	1,054,000	Indiana & Michigan Electric Co.	1972
	Bridgman	1,060,000	Indiana & Michigan Electric Co.	1973
	Midland	530,000	Consumers Power Co.	1974
	Midland	800,000	Consumers Power Co.	1975
MINNESOTA	Elk River	22,000	Rural Cooperative Power Assoc.	1962
	Monticello	471,700	Northern States Power Co.	1970
	Red Wing No. 1	530,000	Northern States Power Co.	1972
	Red Wing No. 2	530,000	Northern States Power Co.	1974
NEBRASKA	Fort Calhoun	457,400	Omaha Public Power District	1971
	Brownville	778,000	Consumers Public Power District and Iowa Power and Light Co.	1972

*Site not selected.

Figure 2-36. Nuclear power plants in the United States.

	SITE	CAPACITY (Kilowatts)	UTILITY	STARTUP
NEW HAMPSHIRE	Seabrook	860,000	Public Service Co. of N. H.	1974
NEW JERSEY	Toms River	515,000	Jersey Central Power & Light Co.	1968
	Toms River	815,000	Jersey Central Power & Light Co.	1972
	Artificial Island	1,050,000	Public Service Gas and Electric Co. of New Jersey	1971
	Artificial Island	1,050,000	Public Service Gas and Electric Co. of New Jersey	1973
NEW YORK	Indian Point No. 1	265,000	Consolidated Edison Co.	1962
	Indian Point No. 2	873,000	Consolidated Edison Co.	1970
	Indian Point No. 3	965,300	Consolidated Edison Co.	1971
	Scriba	500,000	Niagara Mohawk Power Co.	1968
	Rochester	420,000	Rochester Gas & Electric Co.	1969
	Shoreham	800,000	Long Island Lighting Co.	1975
	Lansing	838,000	New York State Electric & Gas Co.	1973
	*	1,115,000	Consolidated Edison Co.–Orange and Rockland Utilities, Inc.	1973
	Nine Mile Point	815,000	Power Authority of State of N. Y.	1973
NORTH CAROLINA	Southport	821,000	Carolina Power and Light Co.	1973
	Southport	821,000	Carolina Power and Light Co.	1974
	*	821,000	Carolina Power and Light Co.	1976
OHIO	Oak Harbor	800,000	Toledo Edison – Cleveland Electric Illuminating Co.	1974
OREGON	Rainier	1,105,000	Portland General Electric Co.	1974
PENNSYLVANIA	Peach Bottom No.1	40,000	Philadelphia Electric Co.	1966
	Peach Bottom No.2	1,065,000	Philadelphia Electric Co.	1971
	Peach Bottom No.3	1,065,000	Philadelphia Electric Co.	1973
	*	1,065,000	Philadelphia Electric Co.	1975
	*	1,065,000	Philadelphia Electric Co.	1977
	Shippingport	90,000	Duquesne Light Co.	1957
	Shippingport	783,000	Duquesne Light Co.–Ohio Edison Co.	1973
	Three Mile Island	831,000	Metropolitan Edison Co.	1971
	*	1,052,000	Pennsylvania Power and Light	1975
	*	1,052,000	Pennsylvania Power and Light	1977
SOUTH CAROLINA	Hartsville	663,000	Carolina Power & Light Co.	1970
	Lake Keowee No.1	841,100	Duke Power Co.	1971
	Lake Keowee No.2	841,100	Duke Power Co.	1972
	Lake Keowee No.3	841,100	Duke Power Co.	1973
TENNESSEE	Daisy	1,124,000	Tennessee Valley Authority	1973
	Daisy	1,124,000	Tennessee Valley Authority	1973
VERMONT	Vernon	513,900	Vermont Yankee Nuclear Power Corp.–Green Mt. Power Corp.	1970
VIRGINIA	Hog Island	783,000	Virginia Electric & Power Co.	1971
	Hog Island	783,000	Virginia Electric & Power Co.	1972
	Louisa County	800,000	Virginia Electric & Power Co.	1974
WASHINGTON	Richland	790,000	Washington Public Power Supply System	1966
WISCONSIN	Genoa	50,000	Dairyland Power Cooperative	1967
	Two Creeks No.1	454,600	Wisconsin Michigan Power Co.	1970
	Two Creeks No.2	454,600	Wisconsin Michigan Power Co.	1971
	Carlton	527,000	Wisconsin Public Service Co.	1972

Figure 2-37. Nine Mile Point Nuclear Power Plant.

Figure 2-38. Pathfinder Atomic Power Plant.

HUMBOLDT BAY   Eureka, California

Figure 2-39. Humboldt Bay Boiling Water Reactor.

ELK RIVER   Elk River, Minnesota

Figure 2-40. Elk River Reactor.

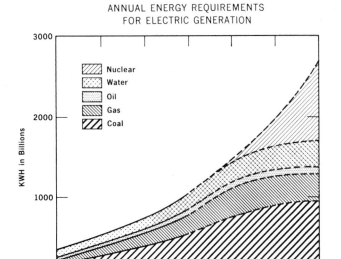

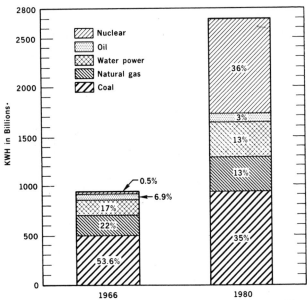

Figure 2-41. U.S. energy requirements through 1980.

Figure 2.36 is a compilation of the principal nuclear electrical power plants in the United States, their start-up dates and their capacity for producing electricity. Photographs of several of these plants are shown in *Figures 2.37* through *2.40.*

The Future of Nuclear Power

At the beginning of 1969, the 13 central station electrical power reactors in operation in the United States with their combined capacity of 2,723.7 megawatts of electricity accounts for about one percent of the 285,000 megawatts total electrical power generating capacity in the United States. But in the United States the electrical producing companies are becoming rapidly and firmly committed to nuclear power. As an example, one large company, the Commonwealth Edison Company, was firmly committed a few months ago to nuclear power. They had about 2,100 megawatts on line and under construction at that time. This company in its last annual report stated that by 1980 it is expected that nearly 75% of the company's total electrical production will come from nuclear energy, compared to only 4% today.

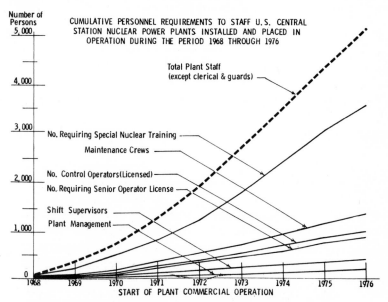

Figure 2-42. *Cumulative personnel requirements to staff nuclear power plant installations in the period 1968 through 1976.*

Figure 2.41 is an estimate of the growth of the annual energy requirements of the United States for electric generation and the contribution which coal, gas, oil, water and nuclear energy sources may make to the total.

Figure 2.42 is an estimate of the personnel requirements to staff central station nuclear power plants installed and placed in operation during the period 1968 through 1976.

The breeder reactor development programme

Widespread use of nuclear power can be reliably projected but to secure the full benefits of the energy of fission it is necessary that breeder reactors be developed.

In breeder reactors excess neutrons, obtained during the operation of the plant for the production of electricity, are used to produce more fissionable materials than is consumed.

When developed for commercial application high gain breeders should increase fuel use of most of the uranium and thorium reserves. High gain breeders should make the nuclear power industry costs relatively insensitive to ore prices.

Interest in the fast breeders is focused on the Liquid Metal-cooled Fast Breeder Reactor (LMFBR) which uses liquid sodium as the coolant. A comprehensive plan for the development of the LMFBR was distributed last year to industry, utilities and government laboratories to achieve further involvement in solving the many difficult problems. The plan includes technical programmes in plant design, component development, instrumentation and control, sodium technology, core design, fuels and materials, fuel re-cycle, physics and safety.

It is hoped that sufficient technology can be developed in the next several years to permit the construction of a demonstration liquid metal-cooled fast breeder reactor at power ratings of about 500 megawatts electrical. Successful operation of such a demonstration unit, could lead to the commercial operation of such breeder reactors in the 1,000 megawatt electrical range in the 1980's.

One preliminary concept of such a LMFBR demonstration plant is shown in *Figures 2.43, 2.44* and *2.45.*

500 Mwe DEMONSTRATION PLANT ARRANGEMENT

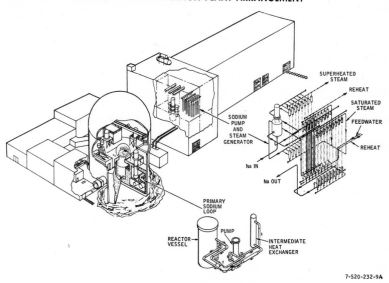

Figure 2-43. Schematic view of a 500 Mwe LMFBR demonstration plant.

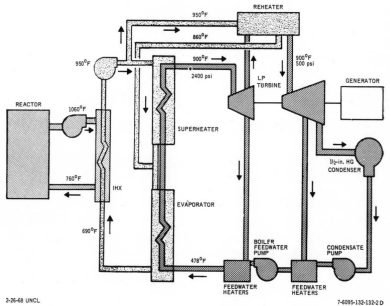

Figure 2-44. Cooling system for LMFBR demonstration plant.

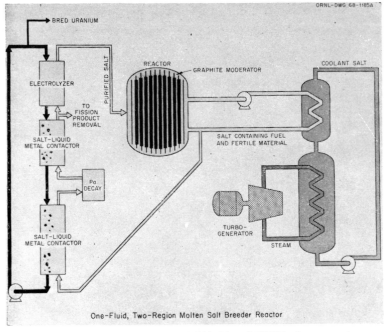

Figure 2-45. Artist's concept of 500 Mwe LMFBR demonstration plant.

Figure 2-46. Design of a molten salt breeder reactor.

The Fast Flux Test Facility (FFTF) reactor, with a design power level of 400 megawatts of heat will provide a versatile test facility in a fast neutron environment typical of that expected in commercial fast breeder reactors. This reactor is scheduled for operation in 1974. It is being built near Richland, Washington. The EBR-2 reactor, now operating at 50 megawatts of heat, is being used to study the behaviour of fuels and materials for the LMFBR programme.

As insurance against possible delays in availability of economically competitive fast breeders, efforts are also continuing on developing advanced converters such as the High Temperature Gas-cooled Reactors and thermal breeders such as the Molten Salt Reactor.

Molten salt breeder

Preliminary studies have shown that the Molten Salt Breeder Reactor shows promise of efficiently using thorium resources. The Molten Salt Reactor Experiment (MSRE) operating on a loading of uranium-233 fuel commenced operation at Oak Ridge in October 1968. This reactor will operate at a power level of 8,000 thermal kilowatts and will provide data needed to evaluate the practicality of molten salt reactors.

The fuel salt used in this reactor is a molten mixture of the fluorides of lithium-7, beryllium, zirconium and uranium. This mixture of salt melts at 813 degrees Fahrenheit and therefore all parts of the reactor that contain the salt must be kept above that temperature. The design of a molten salt breeder reactor is shown schematically in *Figure 2.46*.

CHAPTER 3

Nuclear Energy in Space and Remote Terrestrial Environments

Space Power Systems

Background

Space application scientists have been fortunate to have a well-developed chemical rocket technology at their command. With this technology they have been able to break free from the gravitational forces that bind man and objects to the surface of the earth.

Twelve years ago scientists of the Soviet Union first hurled a small object around the earth. A few months later United States scientists achieved the same objective.

Often since then man and his instruments have ventured into space. Much has been accomplished. Chemical laboratories, soft landed on the moon, have analyzed the material of the lunar surface. Men have gone to the moon and returned safely. Photographs of the Lunar and Martian landscapes hang on many walls throughout the world. We know a lot more than before about the planet Venus from Soviet and American vehicles that have probed that distant planet. Television on a worldwide scale, better weather forecasting, improved navigational aids and a reliable telephonic communication system have all been made possible by satellite observation and relay stations conveniently spaced in appropriate orbits around the earth.

But man is still limited in his exploration and use of space to a great extent by the cost in manpower and effort of providing himself with enough useful energy to enable him to conduct vigorous activities.

For man to survive in space for extended periods he must have enough electrical energy to do many things. He must purify the air and water he needs to stay alive. He must condition the cabin of his spacecraft for livable temperatures and pressures. He must

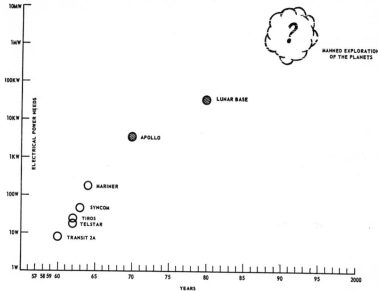

Figure 3-1. Estimate of space power needs.

have enough electrical energy to operate his navigational and communication instruments and to do such chores at his destination as his mission dictates. *Figure 3.1* gives an estimate of how space electric power needs might be expected to grow during the next thirty years as the space exploration programme takes on more ambitious missions. Nuclear energy provides great promise of meeting some of these requirements.

Chemical batteries and fuel cells

Chemical batteries and fuel cells have been used extensively in satellites to provide the relatively small amounts of power required to operate instruments and to perform other functions during many space missions. However, as lifetime requirements go up the weight of the consumable fuel required to be carried along on space trips also goes up rapidly. *Figure 3.2* illustrates how the specific weight requirements for batteries and fuel cells vary as a function of lifetime. It can be seen from this figure that batteries and fuel cells can be used best in those cases where only a few days, or weeks, of operation in space are required.

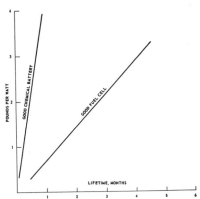

Figure 3-2. Specific weight of batteries and fuel cells as a function of lifetime requirements.

Fuel cells do have admirable properties for space missions whose duration is short enough to permit their use. They are quiet, clean and safe sources of energy. In some cases they can be incorporated into the ecological cycle which keeps the crew alive.

Solar power systems

When electromagnetic radiation falls on a photosensitive surface electrons are set free and if a suitable circuit is provided an electrical current can be produced.

At the surface of the earth the electromagnetic radiation from the sun provides a total of about 150 watts of energy per square foot. Devices can be fabricated that will transform some of the sun's radiant energy into useful electrical energy. Solar cell power systems, being generally of lightweight construction and dependable for relatively long times, are widely used in space applications.

However, the total weight of such solar power systems can become significant. This weight includes the cells themselves, their structural mounts, their protective glass covers, the apparatus required to orient the panels toward the sun and the power system required to operate the orientating apparatus, the secondary batteries and charging devices required to supply power during times of no sunlight, and auxiliary connectors and mounting apparatus.

Provision must also be made for solar cell degradation especially where the system is intended for operation for long times within the high radiation fields often encountered in some space applications. When all these factors are taken into account the weights of practical solar power systems approach about one pound for each watt of electricity produced.

Solar cells require direct access to the local solar energy flux, which varies with the inverse square of the distance to the sun. A solar array which would produce, say 500 watts, in an earth's orbit might be only about 40 square feet in area. At Mars the array would have to be nearly twice as large, at Jupiter about 27 times as large, and at the planet Pluto perhaps fifteen hundred times as large. In the shadow of the moon or of a planet, solar cells of course cannot generate any power.

Atmospheric drag is still another potential disadvantage of the solar power system. The large area required for the collection of the solar energy are subjected to large drag forces when the system passes at high velocity through remnants of the earth's atmosphere. The forces will become especially difficult to deal with when the manned, long-lived, earth orbital stations are contemplated, say in the late 70's.

Present day costs of solar cell systems depend, of course, on the requirement established for the particular space mission involved. As a general rule for systems producing average power of the order of a few watts up to about 200 watts the costs may be about $2,000 to $4,000 per electrical watt.

It is thus clear that solar cells have their limitations. But while they have these disadvantages they do remain most important sources of power in space applications and their use is expected to continue.

Nuclear Electric Power Systems

As man seeks to go farther into space for longer periods, and in more hostile environments, and to do more things while he is there, he will need to take with him compact devices capable of providing him with larger amounts of electrical power; nuclear electric power systems appear to offer promise of meeting these growing needs.

In a nuclear electric power system heat from some nuclear process is transferred from the nuclear source to a power conversion system that converts some of the heat energy into electrical energy.

Nuclear power systems range in complexity from simple radio-isotope generators producing only a few watts of electrical power, now in operation in space, to large reactor installations being planned for future complex space missions.

CONDITIONS UNDER WHICH NUCLEAR POWER SYSTEMS ARE ATTRACTIVE

- **Lack of sunlight**
- **High radiation fields**
- **High power levels**
- **Long life**
- **High atmospheric drag**
- **Heat required in payload**

Figure 3-3. Conditions under which nuclear power systems are attractive.

Figure 3.3 summarizes the characteristics of nuclear power systems which make them attractive relative to other sources of energy for use in space applications.

Isotope	Principal radiation	Half life in years
Thulium - 170	β	0.35
Polonium - 210	α	0.38
Curium - 242	$\alpha\ \eta$	0.45
Cerium - 144	$\beta\ \gamma$	0.78
Promethium - 147	β	2.6
Cobalt - 60	γ	5.3
Curium - 244	$\alpha\ \eta$	18
Strontium - 90	β	28
Cesium - 137	$\beta\ \gamma$	30
Plutonium - 238	α	87.4

Figure 3-4. Some characteristics of isotopes of current interest in space application.

Power from Radioisotopes

Radioisotopes differ in the type of radiation they emit. The radiations are mainly of three types — alpha particles, beta particles and gamma rays. The alpha particle, nucleus of a helium atom, has little penetrating power. Its entire kinetic energy is quickly absorbed by interaction with atoms near the surface of an absorber. The beta particle can penetrate several millimetres of a material like aluminum before its kinetic energy is transformed into heat. Gamma rays are highly penetrating electromagnetic waves similar to X-rays. They can penetrate several centimetres of metal before they are absorbed.

Radioisotopes also differ in the rates at which they decay. Most of the known radioisotopes decay so rapidly that no practical use can be made of them. When all factors are considered, including ease of production, there are only about ten radioisotopes that are of present interest for space applications. *Figure 3.4* lists these isotopes and some of their characteristics.

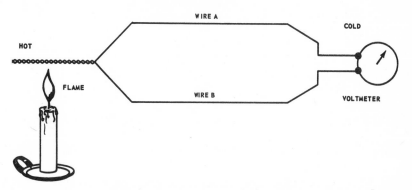

Figure 3-5. The thermoelectric effect.

A German physicist, T. J. Seebeck, noted in 1821 that an electrical voltage appeared across the terminals of two wires made from dissimilar metals. The principle is illustrated in *Figure 3.5*. For more than one hundred years the main practical use of the Seebeck Effect was the measurements of temperature differences. It was not considered practical to use the effect in the generation of electrical energy because of the inherent losses in metal wires.

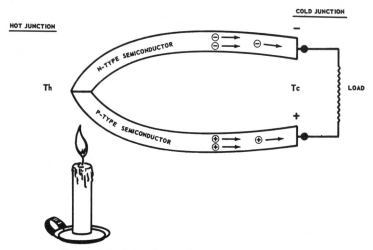

Figure 3-6. The thermoelectric generation of power.

Some materials do not conduct electricity as well as metals but they are not as poor conductors of electricity as are insulators. These materials are called semiconductors. They are of two types, negative (*n*) and positive (*p*), depending on whether they have in effect an excess of negative electrons or of "positive holes". When two pieces of *p* and *n* type semiconducting materials are joined in a way similar to that shown in *Figure 3.6*, heat at the hot junction drives the excess electrons and excess "positive holes" toward the cold junction where a negative and a positive terminal are produced. The larger the temperature difference, $T_h - T_c$, the larger the voltage difference.

With the development of semiconductor materials like lead-telluride (PbTe) and silicon-germanium (SiGe) it became practical to convert heat energy directly into electrical energy by using the thermoelectric effect.

Of course any source of heat can be used to power a thermo-electric generator. The Soviets have manufactured a power generator for use in the bush country from a kerosene lamp with some thermoelectric elements placed in the chimney. Other thermoelectric generators are commercially available in the United

States. These use bottled gas as fuel. They can furnish a few watts of electricity and have obvious uses in remote locations.

It is never possible to transform the heat applied to a thermo-electric element into electrical energy with one hundred per cent efficiency. A French engineer, Sadi Carnot, gave a great deal of consideration to the efficiency of an ideal heat engine. He reasoned that the efficiency, E, of such an engine would be given by the expression

$$E = \frac{T_h - T_c}{T_h} \tag{1}$$

where T_h is the temperature of the heat source in
degrees Kelvin, and
T_c is the temperature of the waste heat reservoir
in degrees Kelvin.

Of course an actual heat engine, like a real thermocouple, is less efficient than the ideal one but equation (1) shows us that an actual heat engine becomes more efficient as the difference between the input temperature and the output temperature becomes greater.

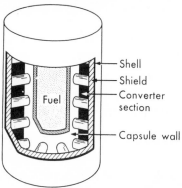

Figure 3-7. Cutaway view of radioisotope generator.

To construct a radioisotope generator, see *Figure 3.7*, one needs only to assemble a radiation source and an array of thermocouples in some convenient form. Thermocouple pairs are usually connected in series electrically and in parallel thermally to provide a high usable output. For protection one must have a radiation shield to stop all stray radiation and a shell of some type to serve as a heat radiator to reject the waste heat.

Figure 3-8. Engineers preparing to launch SNAP-3A.

SNAP systems

The acyronyn SNAP is used in the United States for devices for the production of electrical power from nuclear sources. The name SNAP is derived from the initial letters of the phrase; Systems for Nuclear Auxiliary Power. All odd numbered SNAP power systems, like SNAP-1, SNAP-7, SNAP-19, and SNAP-27, use the energy of decaying radioactive isotopes as their primary heat sources. The even numbered SNAP systems, like SNAP-2, SNAP-8, and SNAP-10A, use nuclear fission reactors as their primary heat sources.

In the United Kingdom the acyronyn RIPPLE, for Radio Isotope Powered Prolonged Life Equipment is used to designate their radio-isotope thermoelectric power generator.

SNAP-3

The SNAP-3 thermoelectric conversion unit was successfully tested in January 1959. It produced 2·5 watts of electricity from

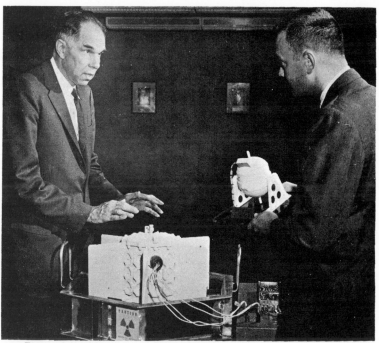

*Figure 3-9. Glenn T. Seaborg, Chairman, U.S. AEC, compares a
SNAP 9-A generator with a model of SNAP-3A held by an
assistant.*

its charge of polonium-210 fuel.   The SNAP-3 was an experimental
model for the SNAP-3A, the first nuclear powered generator used
in space.   *Figure 3.8* is a photograph of engineers at Cape Kennedy
preparing to launch SNAP-3A as a power source in TRANSIT 4-A,
a navigational satellite, on June 29, 1961.   This generator produced
2·8 watts from a plutonium-238 metal fuel element contained in a
rugged capsule mounted in the centre of a sphere of about five
inches in diameter and housed in a white-coated copper shell that
reduced solar energy absorption to a tolerable level and simultaneously allowed the excess heat from the decaying plutonium-238 to
radiate to its environment.   The generator was placed in the
vehicle to provide power to instrumentation and to two of the four
transmitters in the satellite.   The generator, designed for a 5 year
life expectancy, is still functioning.   A similar SNAP-3A unit was
launched in November 1961 as the power unit in TRANSIT 4-B.

SNAP-9A

An improved radioisotope generator, designated SNAP-9A, supplying about 25 watts of electricity had been used by the end of 1963 to power orbiting satellites. This generator weighs about 27 pounds and measures about 20 inches in diameter by about 10 inches high. Metal plutonium-238 was used to fuel this generator. SNAP-9A is designed to have a generator life of about five years. *Figure 3.9* is a photograph showing the general characteristics of SNAP-9A.

Two SNAP-9A units which were launched in September and December 1963 achieved long-lived earth orbits and they have functioned flawlessly although the satellites in which they were used have become inoperative. One SNAP-9A unit was launched in April 1964, but the rocket guidance mechanism failed and the generator re-entered the atmosphere of the Southern Hemisphere and burned up as designed.

SNAP-11

An experimental radioisotope power generator, designated SNAP-11, and originally designed for an unmanned lunar experiment was constructed and tested in 1967.

The unit weighed about 30 pounds and measured about 20 inches by about 12 inches. Fuelled with curium-242 it would have supplied from 21 watts to 25 watts depending on whether it were exposed to sunlight or to darkness on the moon.

The generator was designed as a back-up to a solar cell power supply which was selected as the power unit on the SURVEYOR lunar probes.

SNAP-19

The SNAP-19 thermoelectric generator consists of two 25 watt modular units and is based on SNAP-9A technology. SNAP-19 was developed for use on the NIMBUS B weather satellite. The NIMBUS B satellite is launched in circular polar orbits and observes the earth's atmosphere from an altitude of abut 500 miles.

SNAP-19 is fuelled with a capsule of plutonium-238 in the centre of the generator. The fuel is enclosed in a high strength cobalt-nickel alloy case and graphite capsule. The energy from the decaying plutonium heats up the inner face of an array of lead-

SNAP-19 GENERATOR

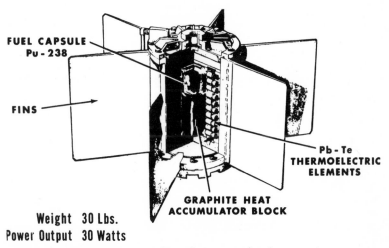

FUEL CAPSULE
Pu - 238

FINS

Pb - Te
THERMOELECTRIC
ELEMENTS

GRAPHITE HEAT
ACCUMULATOR BLOCK

Weight   30 Lbs.
Power Output   30 Watts

Figure 3-10. SNAP-19 radioisotope electric generator.

telluride thermoelectric elements.   The outer face of the array is
cooled by radiation fins.   See *Figure 3.10* for a cutaway drawing of
a SNAP-19 Radioisotope Electric Generator.

The capsule and its contents are designed to withstand ocean
impact and sea water immersion.   A defect in the first stage rocket
booster during a May 1968 launch caused the Range Safety Officer
to destroy the spacecraft.   The satellite containing the SNAP-19
fuel capsule fell in the Pacific Ocean off the California Coast.   The
submerged satellite was detected after an underwater search and the
two valuable plutonium-238 capsules were recovered.   The capsules
suffered no apparent damage.   The NIMBUS B weather satellite is
rescheduled for launch in 1969.   Again a SNAP-19 isotope
generator will be used.

SNAP-27

The development of SNAP-27, a radioisotope thermoelectric
generator, was originally initiated as an alternate source of power to
serve the Apollo Lunar Surface Experiments Package (ALSEP)
which an Apollo astronaut would deploy after landing on the moon.

The SNAP-27 generator is intended to provide a minimum of 63 watts of power for at least a year. The total weight of a SNAP-27 unit is about 66 pounds. A cutaway sketch of a SNAP-27 generator is shown in *Figure 3.11*

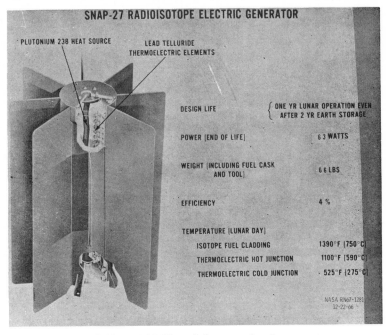

Figure 3-11. SNAP-27 radioisotope electric generator.

SNAP-27 units, like the SNAP-19 ones, are fuelled by plutonium-238 and were specially designed for use on the lunar surface. The lead-telluride thermocouple elements during the lunar day are expected to reach about 525°F at the cold junction and about 1100°F at the hot junction. The output voltage is 14 volts DC and two circuits, each carrying two amperes of current, can be serviced.

Five SNAP-27 generators were delivered by the AEC to NASA in 1968 for Apollo flights and for follow-on missions. Over 25,000 ground test hours have been logged by three generators to demonstrate the long-term capabilities of these units.

TERRESTRIAL APPLICATIONS

While space applications for SNAP devices continue to be most glamorous there are many places on earth where electrical power is needed in modest quantities for important uses. Radioisotope power units have helped solve some of the problems of remote weather stations, of surface and ocean-bottom beacons for navigational purposes and for other special installations at remote desolate spots.

The world's first atomic powered weather station was placed in operation on August 21, 1959, on Axel Heilberg Island in Canada, about 700 miles from the North Pole. Axel Heilberg Island is a bleak, uninhabited land which can be reached only with a modern ice-breaker at the peak of the summer thaw. The generator produced a continuous output of five watts from an array of 60 pairs of lead-telluride thermocouple elements arranged around a cylindrical strontium-90 source. Electrical energy was used to charge chemical batteries between scheduled times for data transmission. Every three hours the stations transmitted data on temperature, wind, and barometric pressure to a manned station at Resolute, 575 miles to the south. Excess heat from the generator was used to keep the data collecting and transmitting equipment within a narrow temperature range.

The unmanned weather station at Axel Heilberg proved to be a complete success. During the two-year period of the test atomic

SOME APPLICATIONS FOR SNAP-7 DEVICES

Designa-tion	Use	Power (watts)	Size (inches)	Weight (pounds)
7A	Navigational Buoy, Baltimore, Md.	10	20 x 21	1870
7B	Lighthouse Tower, Chesapeake Bay, later in Gulf of Mexico as power source for off-shore oil platform	60	22 x 35	4600
7C	Antarctic Weather Station	10	20 x 21	1870
7D	Floating Weather Station	60	22 x 35	4600
&E	Undersea Navigation Beacon	7.5	20 x 21	600

Figure 3-12. Some applications for SNAP-7 devices.

Figure 3-13. SNAP-7A — navigational buoy, Baltimore, Md.

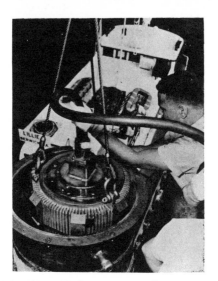

Figure 3-14. SNAP-7B being connected to navigational aids.

Figure 3-15.   SNAP-7B being hoisted inside lighthouse.

power demonstrated that it had many advantages over any un-attended battery operated station.

SNAP-7

The success of the Axel Heilberg Island experiment led to the development and use of several strontium-90 radioisotope generators of the SNAP-7 series. *Figure 3.12* lists some of these applications.

Figure 3.13 is a photograph of SNAP-7A, a floating light buoy which is installed in the Chesapeake Bay near Baltimore, Maryland.

Figure 3.14 gives some details of SNAP-7B, a 60-watt strontium-90 powered thermoelectric device which served for two years without maintenance or service in the Chesapeake Bay as the power source for a lighthouse.   *Figure 3.15* shows the SNAP-7B generator being hoisted inside the Baltimore Lighthouse.   Later, in 1965, it was relocated to provide power to an unmanned Phillips Petroleum Company offshore oil platform, 40 miles southeast of Cameron, Louisiana.   The generator operates flashing navigational lights and in bad weather a fog horn.   *Figure 3.16* shows SNAP-7B in a new location.

Figure 3-16. SNAP-7B operating flashing navigational lights.

Figure 3.17. SNAP-7C being installed near South Pole.

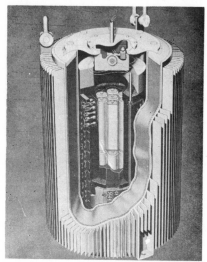

Figure 3-18. Cutaway view of SNAP-7D.

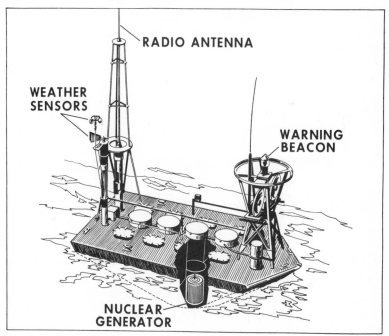

Figure 3-19. Nuclear powered deep-sea weather station.

Figure 3.17 shows the installation of a SNAP-7C radioisotope generator in the ice 700 miles from the South Pole. It provides power for an unattended weather station.

Figure 3.18 shows details of the SNAP-7D generator, which powers the world's first nuclear powered weather buoy, located in the centre of the Gulf of Mexico. Shown in the view are 7 of the 14 tubeless fuel capsules at its centre. Each of the capsules contain strontium titanate providing the source of power. Around the fuel capsules are 120 pairs of lead-telluride thermoelectric elements which convert the heat from the decaying strontium radio-isotope into electricity. A 3-inch shield of depleted uranium prevents radiation from escaping. Double layers of Hasteloy C, a noncorrosive alloy, encase the fuel. The fins on the outside are for cooling. The two wires emerging from the top carry the current from the generator to its points of use. *Figure 3.19* shows how SNAP-7D was installed on a barge 10 ft. × 20 ft. and anchored in 12,000 feet of water.

An experimental strontium-90 powered acoustic navigation beacon, SNAP-7E, now rests on the sea floor in 15,000 feet of water near Bermuda. *Figure 3.20* shows the unit being prepared for shipment.

Figure 3-20. SNAP-7E being prepared for shipment.

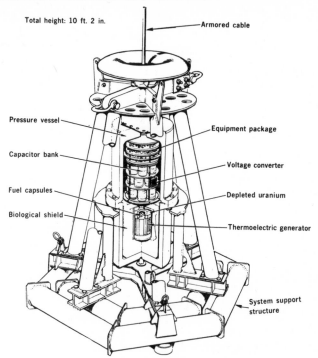

Total height: 10 ft. 2 in.

Armored cable

Pressure vessel

Equipment package

Capacitor bank

Voltage converter

Fuel capsules

Depleted uranium

Biological shield

Thermoelectric generator

System support structure

Figure 3-21.   Cutaway view of SNAP-7E, a strontium-90 powered acoustic navigation beacon.

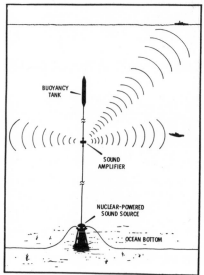

BUOYANCY TANK

SOUND AMPLIFIER

NUCLEAR-POWERED SOUND SOURCE

OCEAN BOTTOM

Figure 3-22.   SNAP-7E implantment.

Figure 3.21 is a cutaway view of the installation. The SNAP-7E isotope generator powers a beacon which produces an acoustic pulse once every 60 seconds. *Figure 3.22* illustrates the use of this device. The sound amplifier is about 3,000 feet above the generator supported by a buoyancy tank 50 feet above it. The lines to the right and left of the generator are grappling cable about 4 miles long which can be used to recover the unit. The unit will provide about 7 watts of continuous electric power for at least two years.

CONTINUING DEVELOPMENTS

The use of isotope power systems in most terrestrial applications is dependent on their economic competitiveness with conventional power sources. The relatively high capital costs of isotopic systems are sometimes balanced by their low operating and maintenance costs and their long lifetime. There are, of course, some applications where isotopic systems may be the only practical means of providing reliable power. Notable among these unique applications of isotopic power are deep-sea, ocean bottom and remote surface installations requiring from 10 to 200 watts of electrical power. In the United States two projects are being conducted for the development of

Figure 3-23. SNAP-21 isotope generator for deep-sea use.

devices for these applications.   These programmes are designated as SNAP-21 and SNAP-23.

SNAP-21

The SNAP-21 systems are being developed for unattended underseas applications.   These units of 10 watt and 20 watt design will use strontium-90 as their fuel.   Structurally sound thermo-electric converters have been built in this power range and stable operation has been obtained in the 1,000°F temperature range. The 10 watt units, which are being developed, measure about 16 by 24 inches and weigh about 500 pounds.   The fuel sources and the thermoelements are enclosed in a corrosion resistant pressure vessel capable of withstanding pressures up to 10,000 pounds per square inch.   The 10 watt units should be ready for ocean testing by the summer of 1969.   The test results will be used to provide design and operational data for the development of the 20 watt and 60 watt units.   *Figure 3.23* is an artist's concept of a SNAP-21 isotope generator for deep sea use.

SNAP-23

The SNAP-23 development programme is intended to result in the design of highly-efficient and flexible radioisotope generators in varying power levels from 25 to 100 watts for surface applications.   These units would be fuelled with strontium-90 and have as an objective the production on electricity in the 60 watt unit at a cost of less than $10 per kilowatt hour, the current cost of electricity to hundreds of operators of gas and oil offshore platforms in the Gulf of Mexico.

The SNAP-19 and SNAP-27 units represent just about the present limit of technology for radioisotope thermoelectric devices. Basic to further advances are the achievements of higher operating fuel temperatures and improvements in power conversion technology.

Of course, it is possible to design a radioisotope thermoelectric generator of almost any capacity by designing a generator and engineering the unit for optimum performance.

SNAP-29

Some work has been done on a SNAP-29 device designed to produce several hundred watts of electrical power for periods of

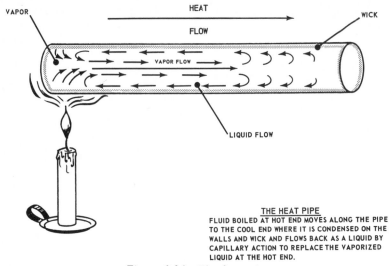

Figure 3-24. The heat pipe.

three or four months in space. The fuel used in this unit would have been polonium-210 which has a 138 day half life. This fuel releases so much heat during the early days of its use that special engineering tricks would have to be introduced to obtain maximum performance.

One of the advanced features which was being tested for SNAP-29 before the work was suspended is the use of heat pipes to transfer heat from the cold junction of the thermocouples to the space radiator surfaces. *Figure 3.24* illustrates the principle of the heat pipe, a most interesting recent technological innovation in heat transfer techniques.

To achieve higher operating temperatures of the thermocouples, it is necessary to develop heat sources capable of higher temperature operation. Studies are underway to improve the performance of existing fuel forms and to develop new fuel forms with better ability to withstand thermal and mechanical shocks, with better thermal conductivity properties and with the ability to retain their good properties when the fuel material undergoes radioactive decay.

Lead-telluride converters operate most efficiently at temperatures in the range 450°F to 1000°F. Silicon-germanium thermoelectrics can operate in the higher temperature range of from 1000°F to 1900°F. By combining the two types of thermoelectrics into an integrated system, it is expected that added efficiency may be attained.

In 1961 thermoelectric generators were capable of operation at about four per cent efficiency. Today they operate with about five per cent efficiency. Perhaps with the improvements sought for in higher temperature fuel sources we can gain another per cent or two in efficiency. Finally, if it is possible to use the two stage integrated thermoelectric generator between the heat source and the radiator one might hope for efficiencies of the order of ten per cent by the end of the 1970s.

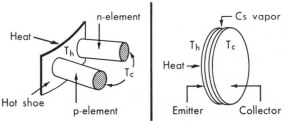

Figure 3-25. Thermoelectric generator elements (left); Thermionic generator elements (right).

Thermionic Devices

A potentially attractive advanced unit being investigated is the isotope thermionic generator. The right side of *Figure 3.25* illustrates the principle of the thermionic energy converter. The principle of operation is quite different from the thermoelectric generators shown on the left side of *Figure 3.25*. Thermoelectric elements are usually made up of pencil-like cyclinders while thermionic converters are usually flat plates separated by a tiny gap filled with a metallic vapour like cesium. The temperature of the heat source for thermionic devices, T_h, must be high enough to boil electrons from the emitter. Therefore the Carnot efficiencies is usually quite high. The electrons, boiled from the emitter, are transported across the narrow gap by the atoms of the cesium vapours which fill the intervening space between the emitter and the collector. An electrical voltage appears in this device between the terminals connected to the emitter and the collector. If the cesium atoms are not present a space charge of electrons would build up in the intervening space to such an extent that the emission of electrons from the emitter would be curtailed. A primary candidate for the fuel for such a device is curium-244. Curium-244 has a half life of 18 years, a reasonably high specific thermal activity, and a relatively long operational life. Studies of this material are one of the components of the advanced isotopes fuels programme.

Availability and costs

The current proven isotope for space power applications is plutonium-238, an alpha particle emitter with a half life of 87 years. In addition to its long life, it is relatively inert chemically, has a high melting point, is quite stable when exposed to chemical, thermal and radiation environments. This isotope is made by irradiating neptunium-237 with neutrons in a fission reactor. Neptunium 237 is a by-product formed in the fuel elements used in nuclear electric power reactors. As the nuclear power industry grows over the next decade or two, it should be possible to recover neptunium-237 from the irradiated fuel elements in sufficient quantities to make a major contribution to the future availability of plutonium-238 for isotope power applications.

In the meanwhile, however, plutonium-238 is in fairly short supply and is quite expensive (about $400 to $700 per thermal watt). Both of the situations assist our scientists and engineers in finding systems to use cheaper materials like strontium-90, a waste product from commercial power reactors.

A large fundamental research programme has been conducted for many years in studying the properties of the transplutonium elements. Much of the information obtained in this fundamental research programme has been directly used in the development of radioisotope power devices. As more elements and isotopes are discovered and as we learn more about the isotopes which have been identified, it is quite likely that isotopic power sources will find wide application.

Dynamic devices

When the radiation from a source of radioactive materials or from a nuclear reactor is absorbed in its environment the heat produced can be used to raise the temperature of some working fluid, like water, to its boiling point. The vapour thus produced

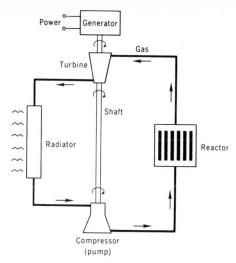

Figure 3-26. The generation of electricity by a dynamic device.

can be expanded as it passes through a turbo-electric generator and returned as a liquid to complete the cycle. See *Figure 3.26.* Other fluids like liquid mercury, potassium, sodium, lithium, or organic materials, or gasses like neon or krypton may be used as the working fluid.

While the turboelectric generator is an efficient device, it involves the operation of high-speed rotating equipment. For useful remote applications, a power generator must be capable of operating reliably without attention for long periods of time. It is therefore now surprising to hear that work was abandoned on SNAP-1, a 500 watt generator using cerium-144 as the fuel, in favour of thermoelectric devices.

Now that the needs for improved technology are becoming more urgent, engineers are turning their attention again to the study of dynamic systems. Two separate gas turbine systems are under study; the Brayton Cycle and the Organic Rankin Cycle.

Brayton Cycle

A gas turbine system, the so-called Brayton Cycle, seems to offer good promise for providing the high efficiency of conversion of heat to electricity needed. In the Brayton Cycle a gas like neon or argon is heated by the decaying radioisotopes. The gas expands through the turbine generating electricity and passes through the radiator where it loses heat to the environment. After this the gas is compressed and returned to the heat source to complete the cycle. The use of an inert gas as the working fluid eliminates chemical corrosion but it is difficult to remove the heat from the expanded gas and a lot of energy is required to compress the gas for return to the system.

Organic Rankin Cycle

An additional power conversion technology, the Organic Rankin Cycle, a two-phase boiling-condensing turbine system may provide high efficiencies at lower temperatures in the several kilowatt range. This cycle could also be used with small reactor systems and can potentially provide efficiencies of the order to twenty per cent.

Nuclear Reactors in Space

Several types of nuclear fission reactor power plants for use in space have been studied in the United States. Some have been tested in terrestrial environment. Others have been flown in space. Most of the space exploration flights up to the present have been in the low-power class. When the space ship engineers were choosing the most appropriate power system for these low-power flights they have usually chosen solar cells, fuel cells, or radioisotope thermoelectric power systems.

Reactor power for space vehicles becomes competitive with solar cells and radioactive power systems when more than a few kilowatts are required for extended periods. It takes between one and two kilowatts per person to keep men alive and comfortable for long periods of time. Certainly if the manned missions of longer than a few weeks are to be successful nuclear power will be required.

SNAP-2

In 1956 work was initiated on the design of SNAP-2, a reactor heated power plant to produce three kilowatts of electricity. *Figures 3.27* and *3.28* are schematic diagrams of the SNAP-2 nuclear power plant and its components.

The heart of the SNAP-2 is the nuclear reactor. In order to minimize the use of valuable uranium-235 and to provide for efficient heat transfer from the core of the reactor to the energy conversion section of the SNAP-2 consists of a bundle of long

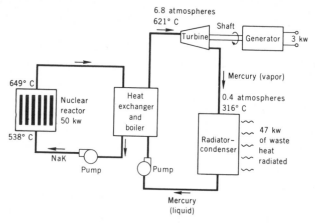

Figure 3-27. *SNAP-2 Schematic diagram.*

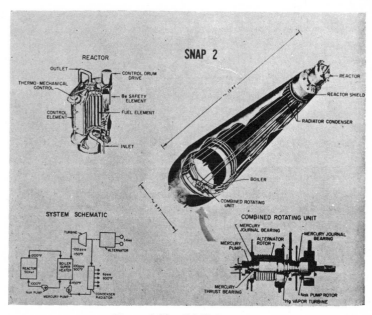

Figure 3-28. *SNAP-2 reactor.*

slender cylindrical fuel elements through which a molten alloy of sodium and potassium metal is pumped. Movable pieces of beryllium core are mounted on the outside of the reactor vessel to reflect neutrons back into the reactor and this serves as a mechanism for maintaining and controlling the power level of the reactor. The fuel elements are made from a mixture of uranium hydride and zirconium hydride and are clad in metal sheaths to protect the contents from the coolant and to prevent the dispersal of the radioactive fission products produced during operation of the reactor.

SNAP-8

In about 1960 development work was begun on SNAP-8, 35 kilowatt nuclear reactor power unit for possible use in orbital laboratories, at a lunar base, in communication satellites, or on deep space missions.

SNAP 8 EXPERIMENTAL REACTOR

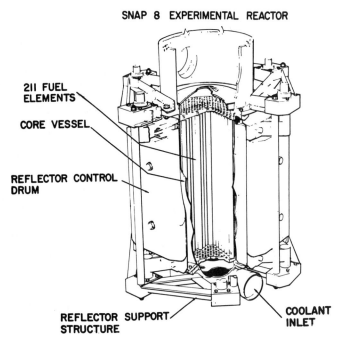

211 FUEL ELEMENTS

CORE VESSEL

REFLECTOR CONTROL DRUM

REFLECTOR SUPPORT STRUCTURE

COOLANT INLET

IO-27-6I

7580-80108

Figure 3-29. SNAP-8 experimental reactor.

In principle SNAP-8 is composed of an uranium-zirconium hydride core and is cooled by sodium-potassium (NaK). The control of SNAP-8, like SNAP-2, is achieved by means of a movable block of beryllium metal which reflects varying numbers of neutrons back into the reactor.

Although no SNAP-8 system has been used in any space missions, ground testing of an advanced model, SNAP-8 Experiment Reactor (S8ER) have been quite extensive. *Figure 3.29* is a cutaway drawing of the SNAP-8 Experiment Reactor. More than one year of reactor operation was attained with coolant outlet temperature of 1300°F. Information obtained during the tests assists the design engineers to further improve components and systems performance. For example, the hot cell examination of the S8ER fuel elements found that 80 per cent of the fuel cladding tubes had cracked. As a result, fabrication of the fuel elements for the SNAP-8 Development Reactor (S8DR) was stopped and an intensive effort initiated to identify the cause of the cracks and to provide a solution.

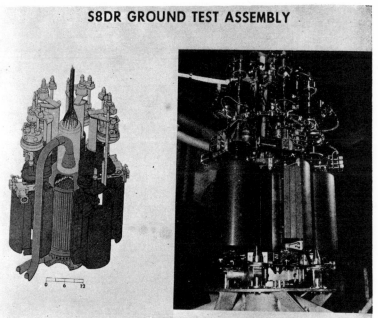

Figure 3-30.   S8DR ground test assembly.

Testing of the next uranium-zirconium-hydride reactor, the SNAP-8 Development Reactor (S8DR), shown in *Figure 3.30*, is in progress. Heat is generated in the core of the reactor by 211 cylindrical fuel elements containing highly enriched uranium mixed with zirconium hydride. The peak temperature in the core is about 1300°F. The core, about two feet high and 2½ inches in diameter, is cooled by pumping a liquid mixture of sodium and potassium through it.

Plans are to operate the test reactor for more than a year at 600 thermal kilowatts at 1300°F. During this period, the reactor will also be operated at power levels up to 1000 kilowatts at lower temperatures. For the test period, none of the heat is converted to electricity as would be done in practical use of the reactor system. Tests will also be conducted on the development of the thermoelectric converter unit including life testing of several thousand hours. It would thus appear that a qualified flight system might be ready for use in a space vehicle by about 1975.

The system could provide power for manned orbiting laboratories —large spacecraft which will undertake research from vantage points high above the earth—and bases on the surface of the moon, from which astronauts can explore the lunar surface. It is being considered for these uses because of its potentially high reliability, small size and long life (two to five years) without need for refuelling or maintenance.

SNAP-10A

The successful flight of the SNAP-10A nuclear power system was an important milestone in the United States space programme. The unit selected for this mission was designed to produce about 500 watts of electricity. The reactor is designed so as to have minimum weight (about 960 pounds) and to be capable of remote start-up in orbit by command from the ground. It is a zirconium hydride reactor fuelled with 4.8 kilograms of uranium-235 and cooled by an alloy of sodium and potassium metals (NaK). A schematic diagram of SNAP-10A is shown in *Figure 3.31*. The NaK leaves the reactor at about 1000°F and is pumped past an array of 2880 silicon-germanium thermoelectric elements and back into the reactor at a temperature of about 900°F.

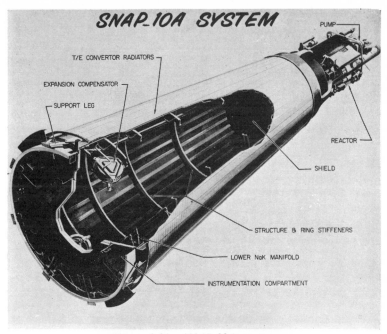

Figure 3-31.   SNAP-10A system.

Figure 3-32.   Launch of SNAP-10A.

The thermocouples are mounted directly between the hot NaK pipes and the surface of a titanium conical radiator which distributes the waste heat to empty space.

The SNAP-10A system was launched on April 3, 1965, at 1:24 p.m. PST by an Agena-Atlas rocket vehicle. *Figure 3.32* shows the launching of this vehicle. Tracking data obtained on the first revolution indicated that the vehicle was in a near circular polar orbit with an apogee of 705 nautical miles and aperigee of 695 nautical miles. The start-up command was given by ground control at 5:05 p.m. The reactor started up without any noteworthy incidents and produced more than 500 watts of electricity successfully and reliably until May 16, 1965, 43 days after launch when an abrupt loss of electrical power occurred. Analyses of the data had indicated that the failure was probably due to events aboard the space vehicle not related to the SNAP-10A system.

Thermionic reactors

To provide for electrical power loads greater than those which

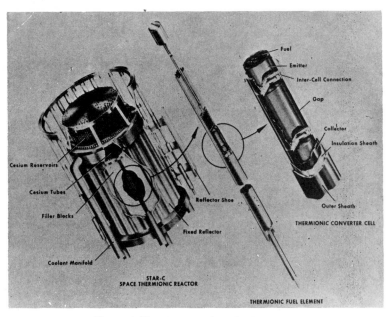

Figure 3-33. A space thermionic reactor.

THERMIONIC CONVERSION

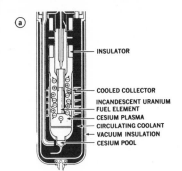
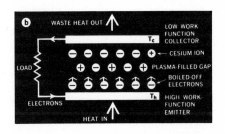

Figure 3-34. Thermionic conversion: (a) Fuel converter diode;
(b) Principle of thermionic conversion.

can be handled by the zirconium hydride reactor system like
SNAP-8, say in the 50 kw to 300 kw range, a more advanced
type of nuclear reactor shows some promise.

This is a reactor in which thermionic generator diodes are
designed as an integral with the reactor fuel elements. *Figure 3.33*
shows a drawing of a compact space reactor of this type. This
reactor would eliminate the need for a turbo-generator in the
separate boiling-condensing thermodynamic loop and efficiency of
the order of 12 per cent have been postulated for the system
provided fuel temperatures of about 3200°F are attained.

In this reactor the combined nuclear fuel-thermionic emitter
element, operating at a high temperature, emits electrons. These
electrons in passing through a space containing hot cesium gas
to the collector form an electric current. The emitter and the
collector form two electrodes of a diode, see *Figure 3.34*, and
must be kept separated by some sort of non-conductor of
electricity. The cesium must be at the optimum pressure to secure
good performance of the diode.

Work is currently in progress in the development of a reliable
thermionic fuel element. The detailed design of a reactor experiment can begin when experimental fuel-emitter elements capable of
achieving very long life are developed. Perhaps the results of
the experiments conducted in the next few years will permit us
to carry out reactor experiments in the 70s.

Nuclear Rocket Propulsion

The amounts of chemical propellants required to launch a few men and their equipment into space is indeed tremendous. Chemical rockets require large amounts of fuel and oxidizers to produce the enormous thrust forces required to accelerate the space vehicle to orbital velocities.

The efficiency of a fuel for rocket propulsion is measured in terms of specific impulse. Specific impulse is defined as the time in seconds during which one pound of fuel will produce one pound of thrust. Thus the larger the specific impulse the better the fuel for thrust development. For chemical fuels the specific impulse is about 400 seconds. It seems clear that one could hope to achieve specific impulse values for nuclear systems at least twice those of the best chemical propellants. In principle this means that twice the payload can be carried on nuclear propelled systems as compared to chemical propelled systems. Because a factor of two in payload increase may be taken in such things as supplies or instruments, the actual effect in space exploration could be much more marked.

As with nuclear electric power the potential advantage of a nuclear propelled rocket lies in the large amount of energy which can be obtained from a small amount of fuel. The nuclear rocket would be designed so that the heat produced in a fission reactor would be used to increase the temperature of liquid hydrogen from —420°F to about 4000°F. The heated hydrogen produces thrust

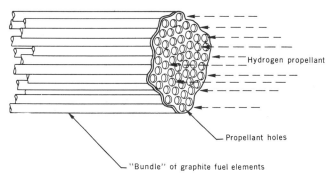

Figure 3-35. Nuclear rocket fuel element.

when it is accelerated as it escapes through a jet nozzle. *Figure 3.35* is a generalized sketch of a typical rocket fuel element.

KIWI reactor experiments

Development work on nuclear rocket engines was initiated in the United States in 1957. The experiments, appropriately named KIWI[1], took place at the Los Alamos Scientific Laboratory. They involved the construction and testing of a series of nuclear reactors designed to operate at about 1,000 megawatts.

The KIWI-A was fired for five minutes in 1959 using pressurized hydrogen gas as the ejected fluid. It generated 70 megawatts of heat power and temperatures of about 1800°K were reached.

Two more KIWI reactors were tested in 1959 and 1960. These experiments showed that satisfactory coatings of niobium carbide could be developed to protect the graphite used as the structural material for the uranium 235 fuel elements and for other reactor components fabricated of graphite from the high-temperature hydrogen. A cutaway drawing of a KIWI-A reactor is shown in *Figure 3.36.*

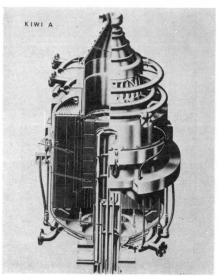

Figure 3-36. KIWI-A reactor.

---

1 New Zealanders know the KIWI to be a tailless, wingless bird which cannot fly.

Figure 3-37. Phoebus 2-A rocket reactor.
Figure 3-38. Phoebus 2-A rocket reactor test.

A new series of more powerful experimental reactors, the KIWI-B experiment, was begun in 1960. The reactors were designed to produce about 1100 megawatts of heat and temperature of the order of 2300°K, sufficient to generate 55,000 pounds of thrust in space. The ejected fluid in the KIWI-B reactors came from a supply of liquid hydrogen. Several successful experiments were carried out which provided the engineers with important design information.

Phoebus reactor experiments

In 1963 it was decided to build and test the Phoebus-2 reactor to provide technology for high-power, high-temperature rocket reactors. *Figure 3.37* is a photograph of the Phoebus-2A reactor on its test stand. On June 26, 1968, the Phoebus-2A reactor operated for 32 minutes. *Figure 3.38* is a photograph taken from over one and one-half miles away. The slender tower to the extreme left of the reactor exhaust plume is over 400 feet high. For about 12 minutes the reactor operated at power levels above 4,000 megawatts. Propellant was allowed to flow through the reactor several times during the run. On July 18, 1968 the Phoebus-2A reactor was restarted for a series of experiments at low and intermediate power levels. On this second run the reactor operated for about 30 minutes at power levels up to 3,700 megawatts.

Figure 3-39. Peewee I test reactor.

Peewee I reactor

Late last year the Peewee I reactor, a unique testbed for fuel elements and support hardware for a flyable space vehicle, was tested. See *Figure 3.39* for a photograph of Peewee I. The reactor operated at significant power levels for 90 minutes achieving an operating temperature of 4,600°K. For more than 40 minutes it produced more than 500 megawatts of power. Although the Peewee reactor is much smaller than the Phoebus-2 reactor — 38 inches in diameter instead of 80 inches — it will permit rapid evaluation of materials and systems for flight qualification. Data obtained during these highly successful tests are very useful in the development of the NERVA flight reactor.

The NERVA engine

In the fall of 1960, the NERVA Programme (Nuclear Engine for Rocket Vehicle Application) began work on the design, development, fabrication and testing of a flight-type version of the 1100 megawatt reactor.

A series of Engineers System Tests on "breadboard" engines

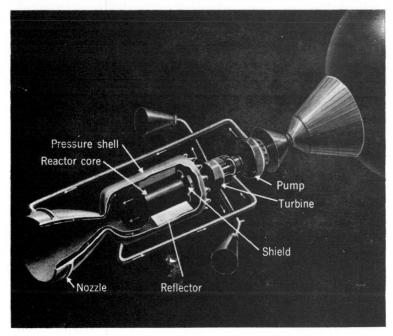

Figure 3-40. The completed NERVA engine.

conducted in 1966 on earthbound units showed that a complete rocket system can start on its own power and operate stably over a wide range of conditions.

By the end of 1966 sufficient information was on hand to give the engineers some confidence that a flyable NERVA engine could be developed. Although a flyable engine has not yet been constructed, *Figure 3.40* is an artist's concept of a complete NERVA engine. *Figure 3.41* is a photograph of a model of an advanced **NERVA** engine intended to produce 60,000 to 75,000 pounds of thrust and approximately 1,500 megawatts of power. The XE Engine, an experimental engine designed to be tested under simulated altitude conditions to approximate the operation of an engine in flight, has undergone sufficient tests to give those working on the NERVA engine confidence that such an engine can be flight-qualified by the mid-70s.

Figure 3-41. A model of the NERVA engine.

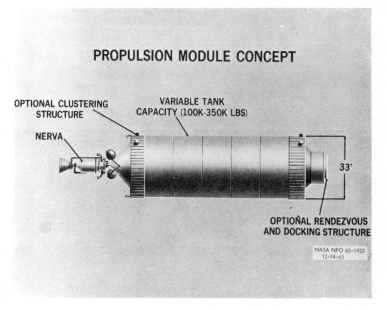

Figure 3-42. The propulsion module.

The propulsion module

The NERVA's thrust of 75,000 pounds is much less than the big chemical rockets like the SATURN V, the key launch vehicle used on the Apollo-8 mission to the moon. Accordingly, the role of the nuclear rocket will probably be as the prime mover to transfer space vehicles from earth orbit into paths toward the moon or planets. Chemical rockets would continue to be used to place the vehicle in earth orbit.

The propulsion module shown in *Figure 3.42* is the present concept of how the NERVA engine would be used. The propulsion module provides the space mission designers with a basic building block that can be used alone or that can be grouped in clusters as required.

Space missions for the atom

The concept of propulsion modules made up of NERVA-like engines seems to be superior to the now dominant chemical rocket for missions where the great economy in propellant consumption is important.

Figure 3.43 is an attempt to compare the payloads which might be carried if chemical and nuclear rockets were used as the third stage in space vehicles.

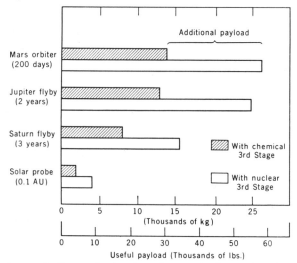

Figure 3-43. Payloads of nuclear and chemical rockets.

CHAPTER 4

Large Scale Industrial and Social Uses for Cheap Nuclear Power

NUCLEAR DESALTING

The need for desalting

Large quantities of pure water are required to meet the needs of modern man. These needs are met by collecting the recently fallen rain and snow from the rivers and lakes of the world and from shallow and deep wells which tap the subsurface sources. In the United States, we presently use about 400,000,000,000 gallons of fresh water each day. This is about 2,000 gallons per person. In many parts of the world pure water continues to be scarce and millions of people today must spend a lot of their energy in searching for water and in conserving their meagre supplies of this vital fluid. Even in a water rich country like the United States, the inhabitants of 1,000 communities must seek fresh water from outside their own areas because their local supplies contain concentrations of salt too high to be consumed.

In 1964, the United Nations surveyed water requirements of 43 countries and found that there are about 20 areas of the world that need desalting and 41 other areas that may need desalting soon.

With the need for fresh water in the United States and elsewhere in mind, the United States is engaged in a vigorous programme of developing desalting technology, especially large-scale nuclear desalting.

Studies have been made in the potential application of desalting technology to the Near East and the rim of the Mediterranean, including Greece, Israel, and United Arab Republic. Nuclear desalting has been considered for Italy, Spain, Mexico, Tunisia and Australia.

If an economically effective way could be devised to get pure water from the ocean on a large scale, the pattern of development throughout the world would be significantly altered. Of course, for a great part of the earth, better water management procedures should provide short-term and intermediate alleviation of water shortages, but there is still considerable motivation in many parts of the world to develop more pure water sources. In London, for example, about 500,000,000 gallons of water are used daily and thrown away. If, say, 75 per cent of this water could be partially purified by conventional sewage treatment followed by partial desalination and recycled, a significant contribution could be made toward prolonging the time before Britain begins to run short of water.

Desalting principles

Sea water contains about 3·5 per cent of mineral contents, largely common salt. There are many methods for reducing this salt content to make the water useful for human consumption and for industrial purposes. All of them depend on an adequate source of low cost energy.

Desalting processes for purifying sea water fall into two classes. First those that take the water away and leave behind the concentrated brine — for example, distillation, freeze separation, solvent extraction, and reverse osmosis processes. In the second category are those processes that remove salt and leave fresh water behind such as electrodialysis and ion exchange.

MULTISTAGE FLASH DISTILLATION

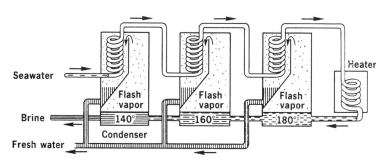

Figure 4-1. Multistage flash distillation plant.

Multistage flash distillation

Distillation is the principal means of purifying saline water. Of the several variants the multistage flash distillation process is the most widely used. In this process hot sea water enters a chamber where the reduced pressure causes it to boil immediately. The steam is then condensed. This operation is repeated successively in a series of chambers at progressively higher vacuum and lower temperatures. *Figure 4.1* shows the principle of a multistage flash distillation plant. *Figure 4.2* is a photograph of a plant using this principle which was originally tested in San Diego and is now operating at Guantanomo Bay, Cuba. The plant during its test operation in San Diego produced 1.4 million gallons of water per day at about $1.00 per thousand gallons.

If large quantities of pure water, say from 100,000 to 100,000,000 gallons per day are required, then this process will probably be the chosen one for most installations, if the choice were made on the basis of present information.

Plant units of about 10 million gallons per day of purified water usually have between twenty and forty stages. Allowing for heat losses as much as 10 pounds of water can be produced in these units from one pound of steam.

Figure 4-2. Multistage flash distillation plant.

LONG-TUBE VERTICAL MULTIPLE-EFFECT DISTILLATION

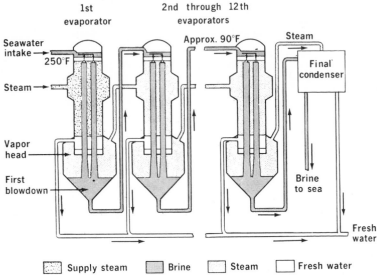

Figure 4-3. Multiple-effect distillation schematic drawing.

For each 1,000 gallons of water produced, the heat consumption may amount to only about 350 kilowatt hours. This heat can be obtained from low grade steam at a temperature of about 260°F. Steam at this temperature is not of much use in the generation of electricity. It is obvious how nuclear heated steam, after passing through turbogenerators for the production of electric power, could be used to power a multistage flash distillation water purification plant. Accordingly, power for operation of a desalting plant might come from the otherwise waste heat of a nuclear power plant.

Multiple-effect distillation

Another type of plant uses the multiple-effect distillation process. In this the sea water passes through a bundle of tubes in a series of evaporators under progressively reduced pressure. Such a plant and a schematic drawing of the progress are shown in *Figures 4.3* and *4.4*. This plant in Freeport, Texas, has a million gallon per day capacity and the process seems to be adaptable to larger units.

Figure 4-4. Long-tube vertical multiple-effect distillation process.

Distillation with vapour compression

If saturated steam is compressed adiabatically, i.e., without a loss or gain of any heat, its initial temperature, t_1, will rise to a value t_2. If this steam is then condensed at a temperature, t_3, (somewhere between t_1 and t_2) slightly more than the original heat of evaporation is recovered at the temperature t_3.   This heat can be used to repeat the original evaporation process to produce the steam at temperature t_1.   Thus water can be evaporated with the expenditure of only the work required for the compression.   This need be only about 3 or 4 per cent of the heat of evaporation.

Figure 4.5 illustrates this principle which has been in use in a demonstration plant at Roswell, New Mexico, where brackish water is converted to potable water. *Figure 4.6* is a picture of the Roswell plant.   This process has some advantages over flash distillation methods on a small scale.   For example, the equipment required is

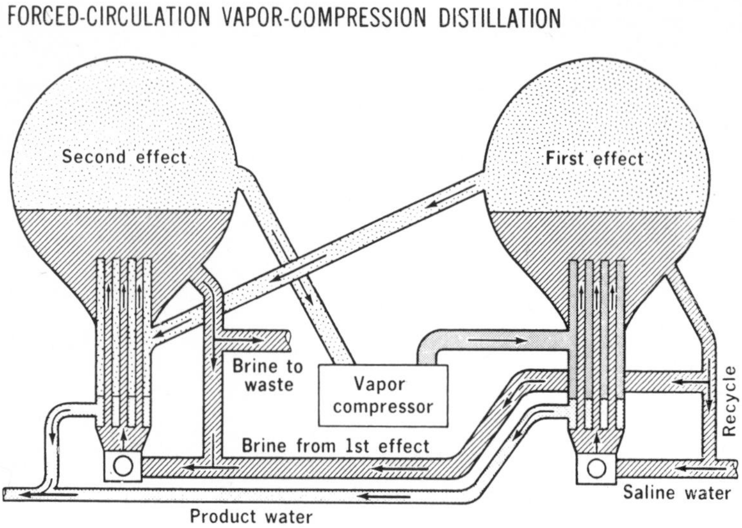

FORCED-CIRCULATION VAPOR-COMPRESSION DISTILLATION

Figure 4-5. Vapor-compression distillation schematic drawing.

smaller. It is not clear, however, whether this process can compete economically with flash distillation on a large scale.

Freezing processes

Because ice crystals cannot form along with the salt impurities without severe local strain, the salt impurities are ejected by the advancing surface of a forming ice crystal. Here then is a way that one can use to make pure water from sea water in a single operation. In this process a "mush" of ice crystals frozen from sea water is scraped from the freezing surfaces and washed free from the adhering brine by recycling some of the purified water. The water-ice slurry is then moved to a melter where the latent heat of the compressed water vapour is used to melt the ice. *Figure 4.7* is a schematic drawing of the Freeze Separation Process. It requires about 540 calories of heat to convert a gram of liquid water into steam. Only 80 calories are required to melt a gram of ice. Thus, since less energy is required to freeze sea water than to evaporate an equivalent amount, the freezing process for desalting water has some potential advantages.

Figure 4-6.   Forced-circulation vapor-compression distillation.

FREEZE SEPARATION

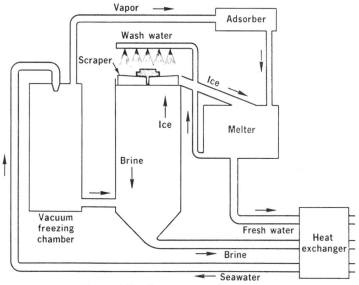

Figure 4-7.   Freeze separation process.

ELECTRODIALYSIS

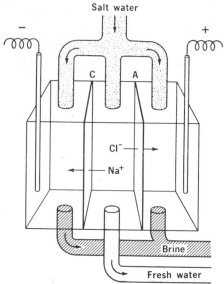

Figure 4-8. Electrodialysis.

Electrodialysis

Electrodialysis is the only commercially used desalting method that removes the minor component, salt, from the major component, water, rather than removing the water from the salt. Electrodialysis makes use of the transport of ions in solution in an electrical field. The equipment, see *Figure 4.8*, consists of a sandwich of alternating cation-and anion-permeable membranes. Upon the application of an electric current, the positive ions, e.g., sodium, pass through the cation-permeable membranes. The negative ions, e.g., chloride, pass through the anion-permeable membrane leaving purer water in the centre between the sandwich.

The Office of Saline Water of the Department of the Interior operates a 250,000 gallon per day electrodialysis desalting plant in Webster, South Dakota. This unit is shown in *Figure 4.9*.

Desalting plants

There are many desalting plants in operation throughout the world. The city of Coalinga, California, now operates its own

Figure 4-9. Electrodialysis desalting plant, Webster, South Dakota.

28,000 gallon per day electrodialysis plant at a cost of $1.45 per thousand gallons. This is enough to provide each resident with 5 gallons per day for drinking and cooking. Previously they had imported fresh water by railway tank car for $7.00 per thousand gallons.

The city of Buckeye, Arizona, supplies all its water needs by its own electrodialysis unit which turns out 650,000 gallons of fresh water daily at a cost of about 60 cents per thousand gallons.

By early 1968 more than 220,000,000 gallons per day of fresh water were being produced in the United States from many small desalting plants. The world's largest single unit plant, with a capacity of 2,600,000 gallons per day began operation in 1967 at Key West, Florida.

In Kuwait there are several government-owned desalting units at one site with a total capacity of 10,000,000 gallons per day. Plants of more than one million gallons per day are also operating in Saudi Arabia, Holland, Italy, The Canary Islands, Curacao, Aruba, on the Persian Gulf, Malta, and Mexico.

The USSR now has a 1,500,000 gallon per day desalting unit that converts brackish water at Shevchenko, a new city on the arid eastern shore of the Caspian Sea. Construction of a dual purpose plant using nuclear heat is also under way there. This plant will supply 150 megawatts of electricity and at the same time produce 150,000 tons of fresh water per day. The plant site is 3.5 kilometres from the shore. The reactor for the plant is a fast breeder reactor and has been tested at the factory. Other potential sites for nuclear desalting plants in the USSR are in the Donets River Basin and in arid parts of Central Kazahstan.

The ideal desalting plant would have low construction, operating and maintenance costs, and would require a minimum of energy input. Any actual plant which might be constructed and operated would be chosen on the basis of the local conditions, on the kind and amount of source water available, and on the required quantities and purity of the output water. In some places, the costs of desalting water are of relatively little consequence. For example, the manufacture of fresh water from sea water on ships at sea is of such paramount importance that the relatively high shipboard costs for distillation units are acceptable.

Economics

The desalting process requires a lot of electrical and heat energy. Electrical energy is required to operate the pumps for bringing in the raw salty water and for distributing the product. Electricity is also required for the general operation of the facility. Heat energy, in the form of steam, can be used to power the separation process.

The design of the desalting part of the plant is not greatly affected by the source of the fuel used to produce the electricity. However, the cost of producing the steam is a major factor in the economic choice between fossil fuels and nuclear fuel. Small size plants are probably best supplied with the required steam by use of conventional fossil fired steam boilers. Large size desalting plants may be more economical to run if they are supplied with steam from nuclear fuel.

If the desalting plant is large enough it is possible to design economical dual-purpose nuclear-fuelled plants. In these plants, high-pressure high-temperature steam produced in a nuclear reactor would first be used to drive turbogenerators for the generation of electricity. The heat contained in the low pressure steam at about 250°F which would be rejected in single purpose electrical power plants would be used to desalt the water.

The economics of nuclear power generation, either steam or electricity, are such that extremely low cost energy can only be produced in very large installations. This has resulted in consideration being given to some large and ambitious projects. Two such possible projects are the Texcoco Project in Mexico City, Mexico, and the Gulf of California Project in Mexico and the United States.

Texcoco project

Modern Mexico City lies in a basin in the south-west corner of a plateau more than a mile above sea level. Water is scarce in that part of the world. Six million inhabitants now draw more than one third of their water needs from 2,000 wells from underneath the city itself.

As the water is pumped from under the city, the water table lowers. The spongy clay subsoil contracts as it dries out. Consequently, the city is sinking at an average rate of about one foot per year. The sinking is not uniform throughout the city and consequently there is considerable damage especially to supply lines, sewage pipes, and tall buildings. An estimate by the Department of Hydrology of Mexico City calculated the cost of the sinking at about 50 million dollars annually just to repair the fresh water and sewer systems and the public buildings. The damage to private buildings was not included in the estimate. Obviously this is an almost disastrous situation.

It has been suggested that salty water extracted from the nearly 450 square mile Texcoco Lake and the subsoil underneath it be desalted for use in Mexico City. This could be done using the heat from a nuclear reactor to power a multiple stage flash evaporator desalting plant. If this can be done, the lowering of the water table under Mexico City would be stopped. Preliminary studies are being

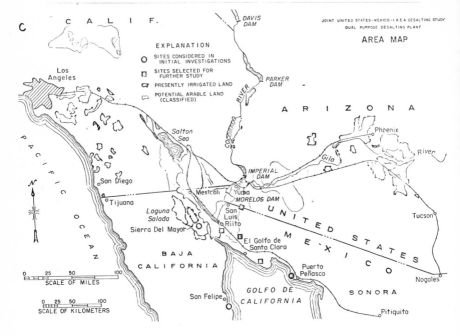

Figure 4-10. Area map — dual purpose desalting plant.

made to determine the feasibility of this great project. It may be the only solution to a tremendous problem.

Gulf of California project

A recent joint study by the United States, Mexico, and the International Atomic Energy Agency provides a good framework for understanding large-scale nuclear desalting.

The southern portions of the States of Arizona and California and the northern portions of the States of Baja California and Sonora, see *Figure 4.10*, have a semi-tropical desert climate and could use large quantities of water. Surface waters are inadequate and subsurface fresh waters are fast being depleted. The land is potentially rich and productive and if the salt water in the nearby Gulf of California could be desalted, the region would prosper with agricultural and industrial growth.

A team of experts about a year ago completed a three-year study of the problems inherent in this situation and made an assessment of the technical and economic feasibility of dual purpose nuclear power and water desalting plants to meet the near and long-term water and power problems of the area.

The team selected desalting plants of 1,000,000,000 gallons per day capacity as the unit size to provide fresh water for this region. These plants would use the vertical tube evaporation process and consume about 2,000 megawatts of electricity. See *Figure 4.11* for an artist's conception of such a basic plant. Several such plants would be required to meet the projected regional demands.

Capital costs for the initial plants were estimated to be in the range of $850,000,000 to $1,200,000,000, depending on site economics. Product water costs as delivered to a major distribution centre could range from about 16 cents to 60 cents per 1,000 gallons. . Electrical power cost estimates varied from 1·8 mils per kilowatt hour to 3·1 mils per kilowatt hour, depending on assumptions made in the fixed charge rate.

Further detailed studies must be made before the true economics of such a large-scale enterprise can be assessed.

Figure 4-11. Nuclear desalting plant, Gulf of California project.

ARID ZONES AND AREAS FACING WATER PROBLEMS

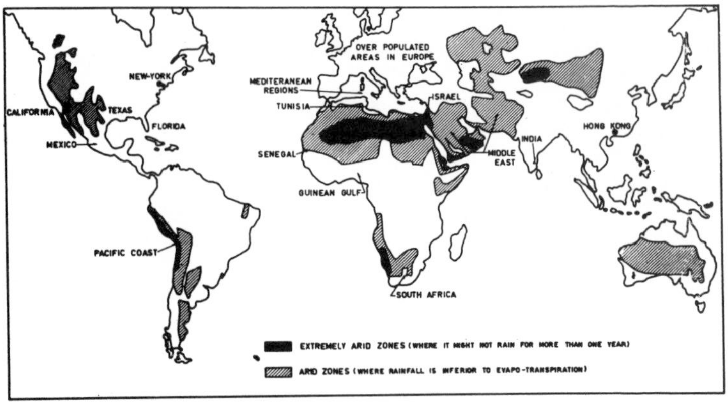

Figure 4-12. Arid zones of the world.

Nuclear Energy Centres

Low-cost energy in large quantities is a prerequisite for industrial development. Many economically sound and well integrated industrial complexes are now clustered around low-cost fossil fuel sources which nature provided. The industrial complex around the low-cost gas fields of Texas and around the low-cost coal fields in South Africa are good examples.

About one-third of the world's land is dry and virtually unoccupied, see *Figure 4.12*. In Australia three-fourths of the continent is in this category. Half of the world's people are jammed into a tenth of the land area. If pure water and cheap power needs can be met, much of the now unused land of the earth could become appealing for human occupancy, new acreage could be reclaimed for intensive cultivation, which would provide food not only for the inhabitants but enough for export to other areas.

Nuclear power has the potential for becoming more economical as the size of reactor plants grow larger. Experience with many nuclear power reactors in the United States indicates that this trend toward lower unit costs with increasing plant size is real. In 1962 the largest U.S. power reactor was 180 megawatts electrical and the largest size on order was 300 megawatts. Most of those on order now are of the order of 1,000 megawatts. This trend is likely to continue as larger fuel fabrication plants and fuel reprocessing plants are constructed and as the trend toward larger energy distribution grids continues.

The history of conventional electric power business is characterized by the increases in the size of the turbogenerators used. Today, units with capacities up to 1,500 megawatts are available. As shown on *Figure 4.13*, it is quite possible that by the year 2000 the unit sizes may approach 5,000 megawatts each. The economies brought about by the operation of large single units together with new developments in breeder reactor technology and electrical distribution techniques, should bring unit costs of nuclear electric power to even lower values than at present.

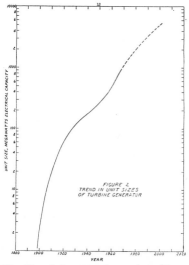

Figure 4-13. Trend in unit sizes of turbine generator.

In perhaps the next 30 years the electrical power needs of some of the larger industrialized countries may be eight or ten times present usage. Then nuclear reactors generating 10,000 to 20,000 megawatts of electricity may become economically attractive. Power cost in such units might be of the order of 1.5 mils per kilowatt hour. At such low power costs a few thousand people can produce food for millions of people. If this should ever happen, a new dimension in man's horizons will unfold with far-reaching social and political implications.

Breeder reactors are now reaching the stage where their designers can build experimental reactors that will require relatively low fuel invention, have high power output, and simultaneously produce more fuel than consumed. In the next 15 or 20 years such reactor systems should be mature enough to begin an era when *inexpensive* and *unlimited* electrical power will be available.

One valuable way to use this large amount of cheap electricity would be to combine advanced cheap nuclear power plants and plants for desalting sea water in a single complex that could then be extended to include other important power consuming industries. Such a complex would, of course, have to be based on the availability of local raw materials, the existence of suitable markets, adequate transportation facilities and the necessary social, economic and political atmosphere for such a large venture.

The Oak Ridge National Laboratory has conducted studies of large seaside industrial and agro-industrial complexes incorporating and integrating a variety of energy-consuming processes and large low-cost nuclear energy centres.

While the Oak Ridge studies suggest even larger units, they speculate about a possible industrial complex centred around a 1,000 megawatt electrical (3,300 megawatt thermal) reactor, probably of the fast breeder type. Such a "NUPLEX" could produce about 250 tons a day each of elemental phosphorus, chlorine, and caustic, 400 tons a day of ammonia, 550 tons a day of oxygen, and still have a surplus of 100 megawatts of electricity for other uses. On the same basis an agro-industrial complex centred around a 2,000 megawatt electrical reactor, a desalting plant with a capacity of 500,000,000 gallons of fresh water per day could produce a variety of crops on about 240,000 acres of surrounding desert and various industrial products. *Figure 4.14* is an artist's conception of a nuclear power agro-industrial complex stretching along the shore of a coastal desert. This NUPLEX is centred about a 10,000 megawatt thermal breeder power reactor. Such a plant would require an investment of the order of $2,000,000,000. This is a large investment in the absolute, but, in terms of the Gross National Product of the earth, it is only a few hours of man's time and effort. Furthermore, 5,000,000 people might be fed by such a plant run by 10,000 people.

NUCLEAR-POWERED AGRO-INDUSTRIAL COMPLEXES

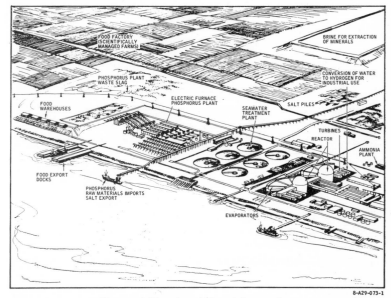

8-A29-073-1

Figure 4-14.   Agro-industrial complex.

A project in the United States recently was given serious consideration for construction.   The Bolsa Island Project, a dual-purpose nuclear station off Southern California, would eventually generate 1,800,000 kilowatts of electricity and distil 150,000,000 gallons of pure water per day.   A photograph of a model of the project is shown in *Figure 4.15*.   While funding of the project has had to be suspended many people are hopeful that solutions can be found to make this project a reality.

Another example of a partial energy centre is a plant in the mid-western part of the United States which will produce both electricity and process steam.   The steam and part of the power would be used by the Dow Chemical Company while the Consumers' Power Company would use the rest of the power in its regular power system.

Plowshare[1]

With the invention of nuclear fission bombs and thermonuclear bombs, mankind was provided with explosive power far greater

1 See Isaiah II:IV

Figure 4-15. The Bolsa Island project — photographs of a model.

than ever before.   Soon after the initial application of nuclear bombs to actual warfare and to planning this distasteful human activity, suggestions were made to apply these explosive forces to peaceful pursuits of man rather than to those which result in his destruction.

Nuclear weapons, because of their tremendous explosive power, have brought about new dimensions in the concept of war and the necessity for maintaining the peace.   Likewise, these new explosives are putting a new technology in the hands of the engineer which will allow him to do things heretofore considered impossible.

It is the purpose of the Plowshare Programme of the U.S. Atomic Energy Commission to conduct a research and development programme in the peaceful uses of nuclear explosions for excavation, underground engineering and scientific application.

The enormous force of nuclear explosions allows one to "think big" in planning suitable projects.   Projects involving massive geographical changes, as well as other ambitious activities, have been envisioned.   *Figure 4.16* lists some of the potential uses of nuclear explosives.

POTENTIAL USES FOR NUCLEAR EXPLOSIVES
Creating harbours and lakes where none existed before.
Cutting passes through mountains.
Stripping overburden from deep mineral deposits.
Digging canals.
Altering watersheds.
Constructing underground reservoirs for retention of gas and water.
Stimulating production of natural gas wells.
Controlling underground water movement.
Desalting seawater underground.
Producing steam and power underground.
Creating basic industrial chemicals directly from mineral deposits.
Recovering minerals from deep underground sources.
Freeing vast oil reserves locked in sand and shale formation now uneconomical to exploit.
Producing transplutonium isotopes.
Studying the inner structure of the earth.

Figure 4-16.   Potential uses of nuclear explosives.

Potential peaceful applications of nuclear explosions require the development of special nuclear devices as well as a detailed knowledge of the behaviour of these devices in all sorts of environments. It is the mission of the Plowshare group to develop a technology for using specially designed nuclear explosion devices that will permit some of these projects to be carried out.

Nuclear explosive devices

The basic principles for fabricating a nuclear explosive device are well known. Such devices can be made from materials in which the fission chain reaction is used alone or from material in which both fission and fusion can take place.

Pure fission devices, because of the large quantities of radio-active fission products produced during the explosion, leave more radioactive residue behind than do devices making substantial use of the energy from thermonuclear fusion reactions. On a pound for pound basis, devices containing thermonuclear mixtures, if properly detonated, are considerably more powerful than pure fission devices. Furthermore, thermonuclear materials, *e.g.*, tritium and deuterium are relatively inexpensive to produce.

ENERGY EQUIVALENT OF ONE TON OF TNT

- 4×10^6 British Thermal Units (BTU)
- 10^9 Calories
- 4.2×10^{16} Ergs
- Fission of 1.45×10^{20} nuclei
- 1,200 kilowatt hours

Figure 4-17. Energy equivalent of one ton of TNT.

For all these reasons, as well as others, the development of Plowshare devices has centred around the economical production of large scale explosions with minimal radioactive fallout and residues. One can now produce Plowshare devices of various sizes, ranging from those with the explosive power of 2,000,000 tons of TNT down to units of about 1,000 tons of TNT. *Figure 4.17* is a conversion table listing data of use in understanding the magnitude of these numbers.

COMPARATIVE ENERGY COSTS
Cost Per 10^6 BTU

TNT explosive	$250.00
10 kiloton thermonuclear explosive	8.75
Ammonium nitrate explosive	4.50
2,000 kiloton thermonuclear explosive	0.075

Figure 4-18. Comparative energy costs per 10^6BTU.

Figure 4.18 lists the costs of producing a million British Thermal Units (BTU)* of energy by several methods. An estimate of costs for a potential user of Plowshare devices projects a charge of about $350,000 for 10 kiloton devices and about $600,000 for nuclear explosion of 2,000 kiloton yield.

Nuclear explosion phenomena

A nuclear explosion raises the temperature of material near its centre to tens of millions of degrees, converting them into gases under enormous pressure. These conditions are reached within a millionth of a second of the time of detonation of the device.

Figure 4-19. The 10 KT thermonuclear explosive and related equipment for Project SEDAN being lowered into the drill hole.

* BTU is the amount of heat required to raise the temperature of one pound of water one degree Fahrenheit.

Figure 4-20. Project SEDAN 3 seconds after detonation.

The high temperature, high pressure bubble of gases expands rapidly, cooling down to about a million degrees in about a ten thousandth of a second and the shock wave which results rapidly moves outward into the environment until its energy is dissipated in vaporizing, melting, crushing, and cracking the surrounding materials. Of course long after the explosive forces have diminished below the level where quantities of material are pushed about, an elastic wave, moving with the speed of sound, may travel for many miles.

Whether the Plowshare device is a pure fission device, or a device using a mixture of fission and fusion, there will be both prompt, characteristic, and residual nuclear radiations which may hinder practical application.

The prompt radiation includes neutrons emitted by the fission fragments soon after their formation, and beta rays and gamma rays from the excited fission fragments.

The long-term radiation comes from the fission fragments and the induced radioactivity of the material surrounding the device. Some unburned fissile material, emitting alpha particles, may also contribute to the long-term radiation.

For several years experiments, consistent with international treaty understanding on the use of nuclear explosives, have been carried out in the United States by the Plowshare group.

Figure 4-21. Project SEDAN 25 seconds after detonation.

Figure 4-22. Project SEDAN.

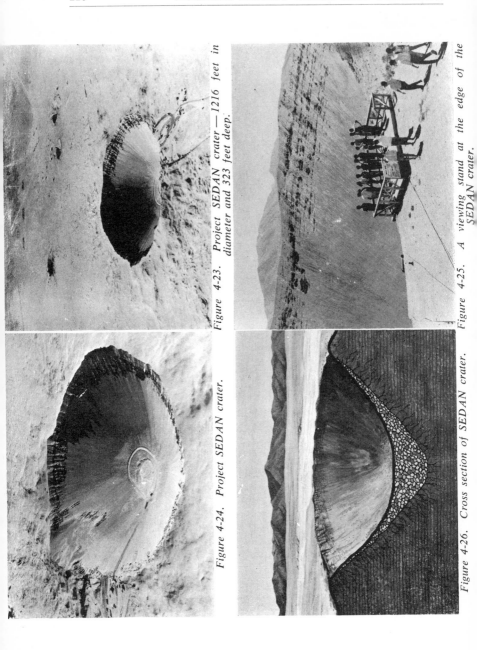

Figure 4-23. Project SEDAN crater — 1216 feet in diameter and 323 feet deep.

Figure 4-24. Project SEDAN crater.

Figure 4-25. A viewing stand at the edge of the SEDAN crater.

Figure 4-26. Cross section of SEDAN crater.

Some of the more important of these large-scale projects will be discussed in this chapter.

SEDAN

The SEDAN experiment of July 6, 1962, is a good example of some of the principles involved in an underground explosion.   In this experiment, a 100 kiloton thermonuclear Plowshare device was detonated 635 feet below the surface in alluvium.   *Figure 4.19* shows the device   being lowered into a 36-inch cased drill hole. Following detonation of the device, the desert rose up to a dome 290 feet as shown in *Figure 4.20* before the explosion broke out. About 12 million tons of rock and earth were lifted up by the explosion with over 8 million tons falling outside the crater as a dense dust cloud, or base surge, that rolled out some two and a half miles.   *Figures 4.21* and *4.22* show two other photographs of the explosion.   *Figures 4.23, 4.24* and *4.25* show several photographs of the crater.   It is the largest excavation ever made by a man-made explosion.

Figure 4.26 is an artist's conception of the cross section of a nuclear crater of the SEDAN type showing the broken rock and fractured area below the floor of the crater.   The actual character-

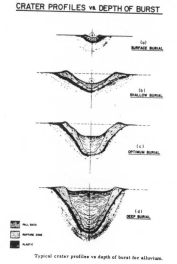

CRATER PROFILES vs. DEPTH OF BURST

Typical crater profiles vs depth of burst for alluvium.

Figure 4-27. Typical crater profiles v. depth of burst for alluvium soil.

istics of craters produced with nuclear explosives vary with the depth of burial of the explosive, the type of surface and subsurface soils, and the type and power of explosive. *Figure 4.27* shows the profiles of typical craters versus the depth of burst for alluvium soils.

DUGOUT

By suitable placement of nuclear explosives in a row and by their simultaneous detonation, a ditch-like crater can be excavated. As a part of the AEC's Plowshare Programme, the simultaneous detonation of five 20-ton charges of a chemical explosive was carried out on June 24, 1964. *Figure 4.28* shows the results of these test explosives in basalt, a hard rock. This view, taken diagonally across the crater, shows the relative difference in height of the lip along the sides (30 to 35 feet) of the crater with the height at the end (10 to 15 feet) shown to the left of the photo.

Figure 4-28. Project DUGOUT crater.

These and other experiments lend strength to the belief that nuclear explosives can be effectively used to construct large canals, say, a sea level canal connecting the Pacific and Atlantic Oceans across the Central American Isthmus.

DUGOUT was a part of a programme to develop the cratering technology required for large-scale nuclear excavations, such as canals, harbours and mountain passes.

Sea level canal

The problems which must be solved before a transisthmian canal can be constructed are, of course, formidable. A very preliminary estimate, made in 1960, for the costs of excavating a canal 1,000 feet wide by 250 feet deep at the centre along these routes ranged from $700,000,000 for the cheapest route to $2,300,000,000 for the most expensive route.

The estimate for the costs for excavating a canal 600 feet wide and 60 feet deep along the same route by conventional excavation techniques ranged from $5,132,000,000 to $13,000,000,000. Whether these cost estimates would stand up under more detailed analysis or not, it does appear that there are real economic advantages to the use of nuclear explosives for such a mammoth under-

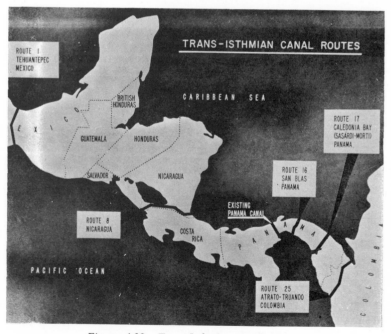

Figure 4-29. Trans-Isthmian canal routes.

taking. Five possible routes have been studied as possible sites for the construction of a sea level canal, with nuclear explosives, across the isthmus. These routes are shown in *Figure 4.29*, a schematic map of this region of the world.

Cape Keraudren Harbour

Considerable attention has recently been given by both the United States and Australia to the concept for the construction of a harbour on the north-west coast of Australia, near Cape Keraudren by the use of nuclear explosives.

The original concept considered the use of a row of five relatively large nuclear explosives detonated with such spacing and at such depths as would produce a harbour about 6,000 feet long by 1,300 to 1,600 feet wide. Side lips would be 200 to 300 feet high. This harbour would be designed to handle ore carriers approximately 1,000 feet long and with a beam of about 135 feet and a draft of about 60 feet.

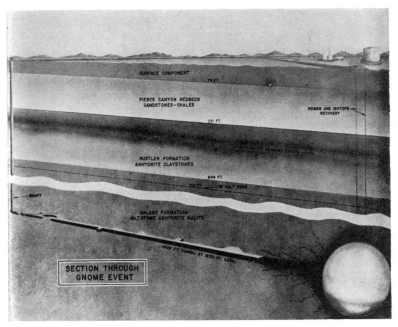

Figure 4-30. Section through GNOME event.

GNOME

In order to study the effect of detonation of a nuclear explosive deep underground, the Project GNOME was conducted in 1961.

In this experiment a nuclear explosive with a yield of 3.1 kiloton was detonated in a salt formation 1,200 feet beneath the earth's surface at Carlsbad, New Mexico. *Figure 4.30* shows an artist's drawing of a section through the geological formation where the explosion took place.

During this explosion, a huge bubble of vaporized material was formed deep underground. This bubble expanded outward rapidly until it produced a cavity with a total volume of about 960,000 cubic feet. In doing this, it melted about 2,400 tons of rock. The melted rock, intimately mixed with about 13,000 tons of salt rock, was hurled back into the cavity by implosion of the walls and, in addition, another 15,000 tons of rock collapsed from

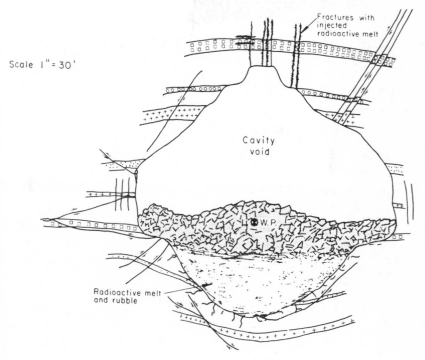

Figure 4-31. Cross section of GNOME cavity.

the roof. Most of the non-gaseous radioactive residue was trapped in the rubble and once-molten salt below the chamber. *Figure 4.31* illustrates the cross-section of the cavity formed by this event. *Figure 4.32* is a photograph of the huge hemispheric cavity measuring 160-170 feet in diameter and 60-80 feet high, which remained after everything was over. The view is toward the top of the cavity that was created by the GNOME shot. Note the stalactites and the size of the man standing in the picture.

Gas stimulation — GAS BUGGY

Natural gas-bearing formations are often so compact in their structure that the gas remains in, and cannot escape from, the rocks. Wells drilled into such formations do not produce gas on an economic basis.

A completely contained underground nuclear explosion, like the GNOME shot, results in a typical sequence of events, such as

Figure 4-32. GNOME cavity—170 feet in diameter, 80 feet high.

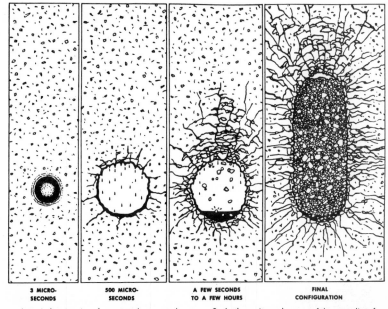

| 3 MICRO-SECONDS | 500 MICRO-SECONDS | A FEW SECONDS TO A FEW HOURS | FINAL CONFIGURATION |

A typical sequence of events when a nuclear explosion is detonated underground, based on data from over 150 AEC experiments. Different geological formations would cause variations in the general outcome.

1. During the first few micro-seconds the explosion creates a spherical cavity filled with hot gases at extremely high pressures.

2. The high pressure forces the cavity to expand. When the pressure inside the cavity is equal to that of the overburden, expansion ceases.

3. As the cavity cools, some of the gases liquefy and the molten rock runs to the bottom. Within a few seconds the cavity roof begins to collapse.

4. Falling rock from the roof creates the chimney of broken rock, which is typical of underground explosions. As the chimney rises to a point where the roof becomes self supporting, its growth ceases. Surrounding the chimney is a broad, highly fractured area which results from the shock of the nuclear explosion.

Figure 4-33. Typical sequence of events when a nuclear explosive is detonated underground.

shown in *Figure 4.33*. Note that in the final configuration the rock formation is often fractured far beyond the physical dimensions of the crater. *Figure 4.34* illustrates how these cracks open up channels in natural gas-bearing rock formation which allow natural gas to be drained off more readily and thus make the gas well more productive.

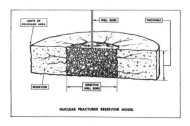 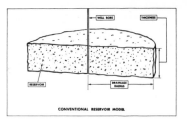

Figure 4-34. *Effect of nuclear fracturing on well-bore of reservoir model.*

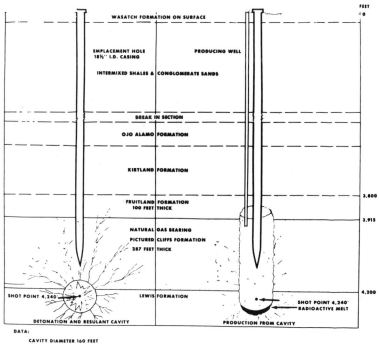

Figure 4-35. *Project GASBUGGY predicted underground effects.*

The first nuclear explosion to test the feasibility of using nuclear explosives to exploit natural resources was detonated on December 10, 1967. The effort was a joint government-industry experiment. The site chosen for the experiment was some 55 miles east of Farmington, New Mexico, in a gas-bearing geological formation of low productivity.

The plan of the GAS BUGGY experiment was to explode a 26 kiloton nuclear explosive at a detonation point 4,240 feet below the surface of the ground. *Figure 4.35* shows a section of the site and the shot point in relation to the geological formations of the area. *Figure 4.36* shows the emplacement and instrumentation plan. *Figure 4.37* shows the general principles involved in gas reservoir stimulation on a large scale.

The real success of the experiment will not be known for a long time. However, preliminary results indicate this technique has increased the gas flow from the formation studied by several times.

In order to determine the effects of nuclear detonation on a gas formation that is geologically different from GAS BUGGY, a second natural gas stimulation experiment was planned. Project RULISON, as this experiment is called, involves the detonation of a 40 kiloton nuclear device 8,400 feet underground about 15 miles from Rifle, Colorado. The gas-bearing Mesaverde formation contains an estimated 8,000,000,000,000 standard cubic feet of natural gas in place, but the reservoir is not commercially productive using conventional techniques. The RULISON shot should determine whether nuclear stimulation of this formation is technically and economically feasible.

Gas storage — KETCH

A joint government-industry study of the feasibility of using a large underground cavity, created by a nuclear explosion of about 25 kilotons about 3,300 feet below the surface to store natural gas, is underway.

Project KETCH would be a straightforward experiment to create such a cavity in a relatively impermeable geological formation. The gas would be pumped into the void during periods of low demand and stored until needed to meet peak load demands. *Figure 4.38* illustrates the principles involved.

Figure 4-36. Emplacement and instrumentation plan for GAS BUGGY.

Figure 4-39. In-situ ore leaching.

TERMINAL GAS STORAGE

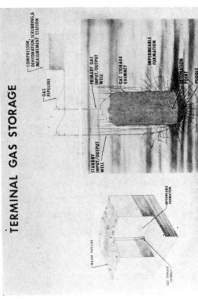

GAS RESERVOIR STIMULATION

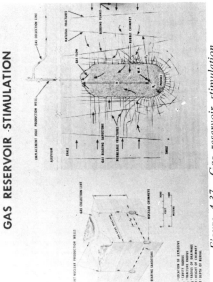

Figure 4-37. Gas reservoir stimulation.

Figure 4-38. Terminal gas storage.

IN-SITU ORE LEACHING

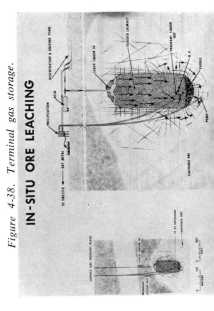

Copper or recovery — SLOOP

Figure 4.39 illustrates the concept being studied jointly by the U.S. Government and the Kennecott Copper Corporation to recover copper from a low-grade copper ore body in Arizona.

In Project SLOOP, as the experiment is called, a nuclear explosive of about 20 kilotons yield would be detonated at a depth of about 1200 feet. Such an explosion should create a chimney of broken rock with a diameter of about 200 feet and a height of about 440 feet and containing about 1,300,000 tons of broken material. The copper would be extracted from this broken rubble by application of leaching technology, a common process used in some copper mining operations.

Oil shale — BRONCO

Oil shale, a fine grained calcareous rock containing solid hydrocarbons, is widely distributed throughout the world and constitutes a major hydrocarbon resource. In Colorado, Wyoming, and Utah vast formations up to 2,000 feet thick occur. Much of this shale yields 25 gallons of oil per ton. When the solid hydrocarbon kerogen is converted by heat to a liquid it gives oil similar to crude petroleum, gas and a carbonaceous residue. It has been estimated that the U.S. reserves amount to about 2,000,000,-000,000 barrels of oil, certainly enough to make it worthwhile to find a way to remove it economically from the shale.

Many people have considered the feasibility of using nuclear explosives to fracture shale so that the oil may be retorted out by combustion of some of the organic matter in the shale. *Figure 4.40* shows the technique which might be used in such a mining operation.

Project BRONCO is being designed to provide some experimental answers to several technical questions related to the recovery of oil from oil shale broken by underground explosions.

CABRIOLET

It is expected that hard rock will be frequently encountered in large scale excavation projects. In order to determine the depth underground at which, according to computation, the best crater would form, a 2·5 kiloton nuclear explosion was detonated in January of 1968.

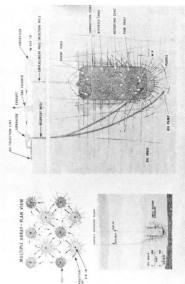

Figure 4-40. In-situ retorting-oil shale.

Figure 4-42. The SCHOONER crater.

Figure 4-41. The CABRIOLET crater.

Figure 4-43. The BUGGY ditch.

crosses the magnetic field at any angle other than at right angles the particle will move in a spiral path along the magnetic field. Positively charged particles will move in one direction and negatively charged particles will move in the other direction. *Figure 5.9* illustrates this point.

Three quantities determine R, the radius of curvature of the path of each particle in a magnetic field: the mass, M, of the particle, the strength, H, of the magnetic field, and the velocity, V, of the particle.

It is helpful to our understanding of magnetic confinement if we make some simple calculations. When an electrically charged ion moves at right angles across a magnetic field the ion experiences a force at right angles to the magnetic field and to the motion of the particle. The magnitude of this force is given by the relationship

$$F_m = He\ V\ dynes \tag{11}$$

Where H is the strength of the magnetic field in gauss, e is the charge in electrostatic units, and V is the velocity of the particle in centimetres per second.

The force F_m tends to bend the particle into a circle of radius, R.

The centripetal force F_c on the particle will be given by

$$F_c = \frac{MV^2}{R} \tag{12}$$

By equating F_c and F_m we have:

$$\frac{MV^2}{R} = HeV \quad ,$$

or

$$R = \frac{MV}{eH} \tag{13}$$

Equation (13) determines the conditions for confinement of charged particles by magnetic fields. Since each particle is constrained to revolve in a circular, or spiral, path of constant radius around the magnetic field lines the particles are not free to migrate to the vessel walls.

There are many shapes of magnetic fields which can confine a plasma. These "magnetic bottles" as they are sometimes called can be classified generally into two classes: (a) the open ended systems and (b) the closed systems. Figure *5.10* illustrates these two types of magnetic bottles.

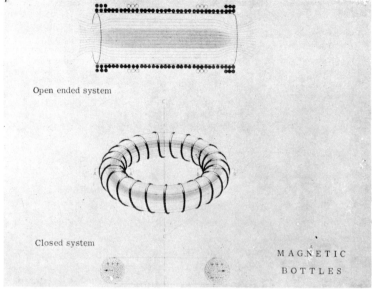

Open ended system

Closed system

MAGNETIC
BOTTLES

Figure 5-10. Magnetic bottles.

Magnetic pressure and plasma pressure balance

An ordinary gas exerts a pressure which can be expressed by

$$P = n k T \qquad (14)$$

where n is the number of molecules per cubic centimetre

k is the Boltzman's Constant and is equal to $1 \cdot 38 \times 10^{-16}$ ergs per degree Kelvin

T is the Temperature of the gas in degrees Kelvin.

In a similar manner the outward pressure, P_o, as a plasma can be represented by

$$P_o = n k T \qquad (15)$$

Here T is the temperature of the plasma in degrees Kelvin, n is the number of charged particles per cubic centimetre and k is Boltzman's Constant.

A magnetic field which confines a plasma must exert an inward pressure on the charged particle of the plasma.

The inward pressure, P_1, on a plasma due to a magnetic field H is given by the expression

$$P_1 = \frac{H^2}{8\pi} \qquad (16)$$

where the magnetic field is expressed in gauss.

In an ideal situation the inward pressure of the confining magnetic field must exactly balance the outward pressure of the plasma, or

$$P_o = P_1 \tag{17}$$

Equation (17) allows us to make an estimate of the magnetic field strength required for confining a thermonuclear plasma and to determine if this is attainable for present methods.

Estimate of magnetic field strength required

A plasma density of about 10^{16} deuterium nuclei per cubic centimetre at an energy of about 10,000 electron volts (about 116,000,000° Kelvin) seems to be attainable with present techniques. This plasma is sufficiently dense and hot enough to be of practical value.

By substituting the above values for the appropriate term of the pressure balance equation we can estimate the approximate strength of the magnetic field required to contain such a deuterium plasma. We can then decide whether such a field is possible of attainment.

From equations (17), (16) and (15) we have:

$$\frac{H^2}{8\pi} = n\,k\,T \tag{18}$$

Putting in the correct values for n, k and T, we have:

$$\frac{H^2}{8\pi} = \left(10^{16}\,\frac{\text{particles}}{Cm^3} \times (1\cdot38 \times 10^{-16})\frac{\text{ergs}}{°K} \times 116,000,000\,\frac{°K}{\text{particle}} \right)$$

Solving for H we find $\qquad H = 60,000$ gauss $\qquad$ (19)

A magnetic field of 60,000 gauss is quite high but it is attainable by present well-known techniques. It thus appears that we shall not meet insurmountable problems from this direction in our search for a feasible method to develop fusion reactors.

The concept of high beta and low beta plasmas

The ratio of the plasma pressure to the magnetic pressure is sometimes denoted by the symbol β and is defined as:

$$\beta = \frac{nkT}{H^2/8\pi} \tag{20}$$

The numerical value to β is often used to describe one of the characteristics of a thermonuclear plasma.

From equation (20) it can be seen that β is proportional to the number of charged particles per cubic centimetre of the plasma or to the density. Now the number of fusion reactions in a hot thermonuclear plasma is proportional to the square of the density. It follows then that the value of β for a thermonuclear plasma is a measure of the number of fusion reactions taking place per unit time. Therefore in the design of fusion reactor systems those with as high a β as practical are preferred because such systems will have more reactions taking place per unit time, and thus produce more power, than systems of low β. On the other hand tenuous plasmas are easier to confine and therefore one must arrive at a judicious compromise in the choice of β for the system proposed for study.

Power generated in thermonuclear plasma

We have seen that it is necessary to keep the particle density in the plasma rather low, say 10^{15} or 10^{16} particles per cubic centimetre, if the plasma pressure is to be kept low enough to be confined by attainable magnetic fields.

Plasmas having this small number of particles per cubic centimetre are quite dilute and generally could be considered as approaching a vacuum. But even at these low densities the rate of energy generation per unit volume from fusion reactions could be of practical interest.

The total energy generated per cubic centimetre per second in a plasma in which fusion reactions are taking place can be computed. The results of such computations for the D-D and D-T reactions are shown in *Figure 5.11*. In this chart the power generated by the D-D and D-T reactions are plotted against the temperature of the plasma in degrees Kelvin.

Power radiation from thermonuclear plasmas

There are several mechanisms by which energy can be lost from the interior of a fusion reactor. One of the dominant ways is by radiation from the plasma itself. When the charged particles collide a type of radiation called "bremsstrahlung" is emitted.

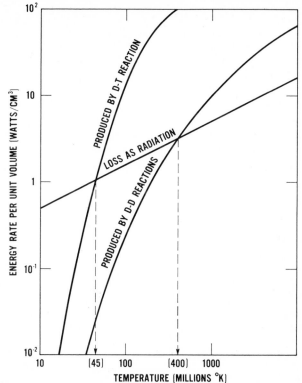

Figure 5-11. Power generated and power radiated by D-D and D-T reactions in a plasma versus plasma temperature at a density of 10^{15} particles per cubic centimeter.

Figure 5.11 shows how the power radiated from a thermonuclear plasma varies with plasma temperature. For a thermonuclear plasma to continue to be a net producer of power clearly it must produce more power than it radiates.

Hence from considering these curves it becomes evident that there is a critical temperature to which a plasma must be heated before the fusion reactions can be self-sustaining. In the case of the D-D reactions this temperature is about $400,000,000°K$. For D-T reaction it is of the order of $45,000,000°K$.

Confinement time

Clearly a hot plasma must be confined in a given region for a time long enough for a large number of fusion reactions to take place. It has been shown that for any deuterium fusion reactor, the product of the particle density, n, in nuclei per cubic centimetre and the confinement time, η, in seconds should be greater than about 10^{16}. For a deuterium-tritium fusion reactor the product of the density and confinement time (η), must be greater than about 10^{-14}.

Thus for a density of 10^{-15} deuterium nuclei per cubic centimetre, the period of confinement must be 10 seconds or more. If the density could be increased to 10^{17} particles per cm³ a confinement time of one-tenth of a second might be sufficient.

To sum up, the following typical conditions seem to be necessary for a controlled thermonuclear reactor:

Temperature	$T \geqq 10^{8°}$ Kelvin
Density	$n \geqq 10^{15}$ nuclei/cm³
Confining Magnetic Field	$H \approx$ 20-50,000 gauss
Confinement Time	$\eta \geqq \dfrac{1}{10}$ second for D-T reactor
	$\eta \geqq 10$ seconds for D-D reactor

Let us turn our attention now to some efforts to achieve these conditions.

Early Experimental Work—The Pinch Effect

As early as 1905, Australian scientists had observed[1] the crushing of hollow copper cylinders by very large electrical currents in lightning discharges. In 1934, W. H. Bennett had predicted that a rapidly moving stream of charged particles would be focused, or constricted because of the influence of the magnetic field which such a current would create. The effect was rediscovered and treated in detail by L. Tonks in 1939.

In the early 1950s scientists in the United States, the United Kingdom, the Soviet Union, and elsewhere, initiated experiments with systems in which a strong electrical current discharge was induced in a gas contained in straight tubes, and in doughnut shaped containers (toroids). In the toroidal devices shown in *Figure 5.12a,* the current produces a magnetic field with the lines of force encircling the plasma that carries the current. The

---

1 "Note on a Hollow Lightning Conductor Crushed by the Discharge", Pollack and Barraclough, Journal and Proceedings of the Royal Society of New South Wales, Vol. 39, p. 131, 1905.

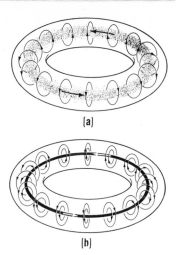

Figure 5-12. Toroidal pinch.

Figure 5-13. Perhapsatron apparatus.

magnetic pressure of the field then "pinches" the plasma into a thin ring near the centre of the tube (*Figure 5.12b*).   In this process the plasma is kept away from the walls of the container and heated by the current itself.   Discharges in straight tubes are similar but simpler.

The pinch effect when first considered seemed to be a simple and promising method for producing and confining a high temperature, high density deuterium plasma.   The first such device, whimsically named the Perhapsatron, consisted of a doughnut-shaped discharge tube a few feet in diameter.   A bank of capacitors was used to store up electrical energy.   The capacitors were then allowed to pulse thousands of amperes of current through the tube in such a way as to create a pinch discharge.   A photograph of the Perhapsatron equipment is shown in *Figure 5.13*. Note the size and scale of this equipment as compared with photograph of equipment used in later experiments.

Enthusiasm for this method diminished when it became apparent that the pinched plasma was highly unstable and would not persist for more than a few millionths of a second — too short to be of practical value.

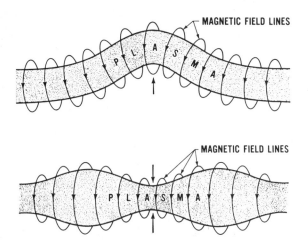

Figure 5-14.   Pinch instabilities.   Upper: kink instability.   Lower: sausage instability.

The upper diagram in *Figure 5.14* illustrates the reason why the plasma in a linear pinch, as such discharges are known, is unstable. The magnetic lines of force remain in a plane perpendicular to the direction of the flowing current. Hence any small kink which might develop in the discharge column causes the magnetic lines of force to crowd more closely together on the concave side of the kink and spread apart on the convex side. The bunching of the magnetic force line creates a higher magnetic pressure locally and, therefore, causes the kink to grow catastrophically.

Operation of the Perhapsatron confirmed the kink instability and thus doomed the linear pinch device as a quick route to thermonuclear reactors.

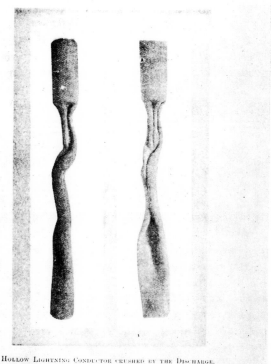

HOLLOW LIGHTNING CONDUCTOR CRUSHED BY THE DISCHARGE.

Figure 5-15. Was this a kink instability in a pinch discharge?
Proc. Royal Society New South Wales, Vol. 39, 1905.

Attempts to stabilize the pinch by heating the plasma more quickly to thermonuclear temperatures resulted in identifying another type of instability. The "sausage-type" instability, shown in the lower diagram of *Figure 5.14* manifests itself as a pinched off region of the plasma wherein much smaller diameters of the plasma develop than in the main body of the plasma stream. The magnetic field is stronger in the constricted region than elsewhere. Consequently the plasma becomes so narrow that it breaks apart.

Figure 5.15 is a reproduction of what may have been the first scientific evidence of the kink instability of an electrical discharge. The original photograph was published here in Australia in 1905.

Attempts were then made to stabilize the pinch discharges against these instabilities. Some of these studies involved the use of an insulated metal ring located in the centre of the toroidal tube. A current in the ring produced an encircling magnetic field. In addition there was an axial magnetic field that surrounded the plasma and was not trapped by it. The plasma was then confined and compressed between these two fields, as shown in *Figure 5.16*.

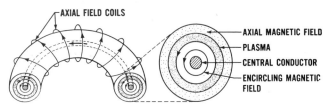

Figure 5-16. Hard core pinch in a torus.

These studies indicated that new types of instabilities were produced. Other experiments involving the use of conducting walls for the discharge tube were carried out. All these studies yielded much valuable information relating to the behaviour of ionized gases, the effects of magnetic fields on plasma configuration, and the general technological problems of high temperature measurements, plasma confinement, and diagnostics. However, the times that the plasmas could be retained in such devices continued to be far too short for thermonuclear fusion power production reactors.

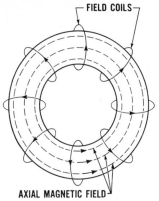

Figure 5-17. Formation of an axial magnetic field in a torus.

The Stellarator Programme

When one confines a plasma in a doughnut shaped tube (torus) and applies an electrical current to field coils as shown in *Figure 5.17,* an axial magnetic field is established in the torus. Note that the spacing of the field coils is closer at the inside of the torus than at the outside. The magnetic field therefore decreases across the diameter of the tube from the inner wall to the outer wall. It is evident that the plasma as a whole will drift toward the outer wall. Confinement of a plasma thus is not possible in simple torus.

However, if the magnetic field within the torus is distorted or twisted in such a way that the lines of force, after making a complete circuit of the tube, do not close on themselves, the plasma drift can be effectively reduced. This is the principle of a device called the Stellarator. The distortion of the magnetic field can be accomplished by twisting the torus into a figure-eight shaped device so that the apparatus no longer is in one plane. Such devices look very much like a pretzel. Some of the early experimental devices were manufactured in this rather expensive and complicated manner. *Figure 5.18* is a photograph of the Model B-3 Stellarator. The figure-eight configuration of the central discharge tube is visible. With this and similar experimental devices, (one was called B-64 because it looked like a "squared 8") scientists at Princeton University gained sufficient knowledge about the principles of stellarators to permit them to initiate design of stellarators with a simple toroidal shape in about 1956. These devices

Figure 5-18. Model B-3 stellarator.

Figure 5-19. Model C stellarator.

used stabilizing coils to provide for the rotation of the magnetic field. Thus the expensive, complex figure-eight devices need not be used.

Figure 5.19 shows the large Model C Stellarator which has been in operation at Princeton University for several years. *Figure 5.20* is a schematic representation of the Model C Stellarator. The magnetic pumping section is used for heating the plasma, and the divertor for removing impurities.

The operation of a stellarator for experimental use is accomplished in three stages. The gas in the tube is first partially ionized by a radiofrequency discharge. The plasma thus produced is confined by the magnetic field and a direct current discharge is then induced from the outside. The ionization of the plasma is greatly increased and its temperature then raised to about a million degrees Kelvin. There are several methods that then can be used in attaining thermonuclear temperatures.

In the magnetic pumping method, magnetic fields are alternatively increased and decreased in rapid succession in a

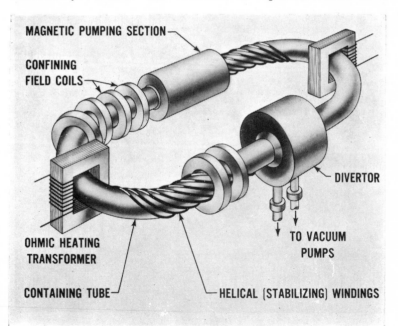

Figure 5-20. The model C stellarator.

portion of the stellarator tube (See *Figure 5.20*). If the correct conditions are attained, the plasma will be heated more on the compression phase than it is cooled on the expansion phase.

The ion cyclotron method of heating involves adjusting the alternating frequency of the local magnetic field to match the frequency at which the nuclei spiral about the main magnetic field. In this way an agitation of the nuclei can result in heating the plasma.

Magnetic Mirror Systems

One can devise a magnetic field about a cylindrical tube in such a way that the field is stronger at the ends than in the middle. *Figure 5.21* is a representation of such a magnetic system. *Figure 5.21a* shows the manner in which the coils are wound; *Figure 5.21b* indicates how the magnetic field strength varies along the tube and *Figure 5.21c* indicates the form of the lines of force which are closer together where the field is stronger. The stronger field at the ends reflect the ions back into the central region much in the manner that ordinary mirrors reflect light. Hence these systems are called Magnetic Mirrors.

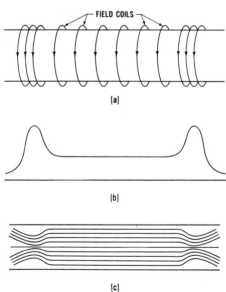

Figure 5-21. A magnetic mirror system.

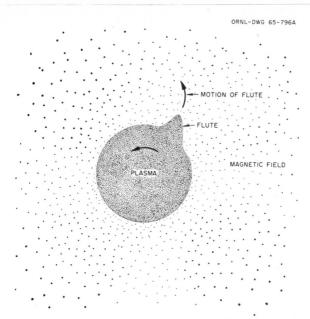

Figure 5-22. The flute instability.

Not all the charged particles in a plasma will undergo reflection at the mirrors. For example, a particle moving along the axis of the vessel will move straight through the mirror. For a particle to be trapped, it must have an appreciable component of its velocity perpendicular to the lines of force in the central region.

Magnetic mirror systems are often operated by generating a plasma in some external manner and injecting it into the tube. Once the plasma is trapped between the mirrors it can be heated by increasing the strength of the magnetic field in the central region, and retaining an appropriate value for the magnetic mirrors.

Many magnetic mirror devices have been constructed for experimentation. A type of instability which manifested itself in these experiments is known as a Flute Instability. *Figure 5.22* shows a cross section view of such a disturbance which extends along the length of the plasma, parallel to the field line, rather like the fluting of a column. This and other types of hydromagnetic instabilities occur as the plasma grows hotter and more dense.

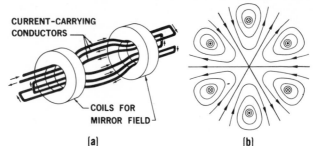

Figure 5-23. Minimum-B configuration.

Methods for stabilizing this type of instability have been worked out. One of them, the so-called Minimum-B configuration (see *Figure 5.23*) is based on the use of conductors parallel to the central axis of the tube in such a way that large electrical currents are in opposite directions through alternate conductors. These conductors are often referred to as "IOFFE BARS", after the Soviet physicist M. S. Ioffe, who first experimented with them.

Plasma injection
Plasmas found in one system can be injected into a second system where they can be trapped and heated.

ALICE
One way to do this is to inject a gas as atoms into the system. Since atoms are electrically neutral they can cross a magnetic field without being affected. Neutral atoms cannot be accelerated to high velocity directly. However, deuterium ions can be accelerated to a few kilovolts energy and a stream of high energy neutral deuterium atoms formed by electrical charge exchange. Once inside the field between the magnetic mirrors the gas can be ionized and converted to a high temperature plamsa. In order to study the problems of Neutral Injection large and complicated equipment like the facility shown in *Figure 5.24* has been found to be necessary. *Figure 5.24* is a photograph of the Adiabatic Low-energy Injection and Capture Experiment (ALICE) equipment at the Lawrence Radiation Laboratory at Livermore, California.

DCX-2
Another method for injecting a plasma into an experimental apparatus involves the generation of a beam of high energy molecular deuterium ions, D_2^+, a molecule of deuterium which

Figure 5-24. The ALICE facility.

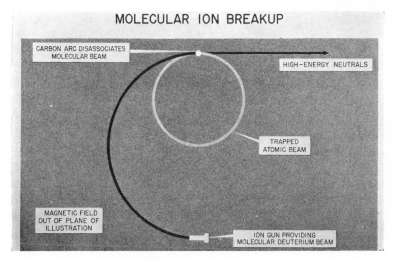

Figure 5-25. Molecular ion breakup.

Figure 5-26.   Direct current experiment No. 2 DCX-2.

has lost one electron. These molecular ions can be disassociated, or broken up, at some point within the magnetic field into a neutral atom of deuterium and a deuterium nuclei. The equation representing this is:

$$D_2{}^+ \longrightarrow D° + D^+ \qquad\qquad (21)$$

A deuterium ion, D^+, produced in this way can be trapped in a magnetic field because its radius of curvature in the field is half that of the molecular ion from which it is formed. *Figure 5.25* illustrates this method for injecting high energy deuterium ions into a magnetic mirror.

Figure 5.26 is a photograph of some of the DCX-2 equipment used at the Oak Ridge National Laboratory in the study of molecular ion injection.

Fast magnetic compression

It is possible to devise a single-turn coil shaped in such a way that when a powerful alternating current passes through the coil that neutral deuterium gas can be successively ionized, heated, compressed and confined. *Figure 5.27* is a representation of such a simple fast magnetic compression system. The coil surrounds a tube containing deuterium gas at low pressure.

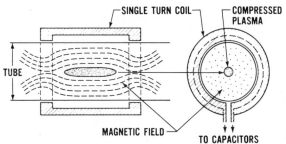

Figure 5-27. *Fast magnetic compression system.*

On the first half-cycle of the oscillations of the magnetic field the gas becomes ionized, on the next half-cycle it is compressed and thereby heated. The Scyllac IV assembly at Los Alamos Scientific Laboratory is the fourth in a series of fast magnetic compression devices. The equipment is shown in *Figure 5.28*.

The Scylla programme has, over the past 12 years, produced some interesting results. It was in Scylla I that the first laboratory thermonuclear plasma was produced in 1958. The Scylla plasmas have been produced which had about the same temperature and radiation properties as the corona of the sun. But the Scylla device is an open pulsed system and is therefore subject to end losses. Consequently, after about five microseconds of operation the plasma is lost from the tube through the ends of the compression coil. These losses make the product, $\eta\rho$, (density times confinement time) down by a factor of 500 to 1000 from the value necessary to show an energy profit.

A new device called SCYLLAC is now being fabricated at Los Alamos and the necessary capacitor bank is being assembled. With this device considerably larger values for $\eta\rho$ should be achieved. The experiment will have a very fast capacitor bank discharging large current through the apparatus. Magnetic fields of the order of 100,000 gauss will be quickly established and maintained for times greater than 100 microseconds. *Figure 5.29* is an artist's drawing of the Scyllac Facility now under construction.

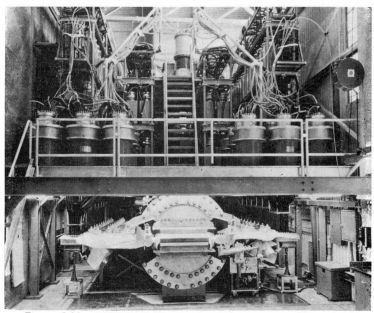

Figure 5-28. SCYLLAC IV experimental apparatus. Upper level —capacitor bank. Lower level—plasma tube and magnetic field compression coil.

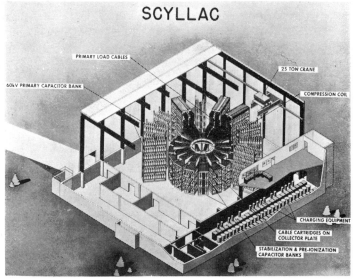

Figure 5-29. SCYLLAC.

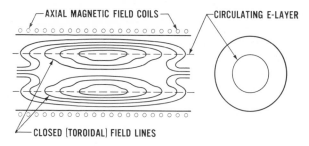

Figure 5-30. The astron concept.

Figure 5-31. The ASTRON facility.

Astron

The Astron system is a concept for producing a closed magnetic field within an open-ended tube. In this experimental device an axial magnetic field is produced in a long cylindrical chamber by passing a current through a coil around the tube in the conventional manner. *Figure 5.30* shows a drawing illustrating the location of these axial magnetic field coils.

An intense beam of high-energy electrons (now about 4 million electron Volts) is injected at one end of the tube. These electrons form a circulating layer about the central axis of the cylinder. The current in this E-layer, as it is called, produces its own magnetic field which interacts with and reverses the axial field yielding a pattern of closed magnetic lines as indicated in *Figure 5.30. Figure 5.31* is a photograph taken from overhead of the Astron device. At the lower right is the 4 million volt linear accelerator which injects an intense (more than 300 amperes) beam of electrons into the long vacuum tank at left. *Figure 5.32* is a side view of the Astron showing the single-layer solenoid 92 feet long built around an aluminum vacuum vessel.

If the E-layer can be built up as envisioned by its designer, the resulting magnetic field should have desirable properties for plasma confinement and should also provide an adequate source for plasma heating.

The Astron experimental device has been studied for about 10 years, significant technological success has been achieved. Electrons have been injected and partially trapped in the apparatus. 300 amperes of 4 Mev electrons have been stored in the E-layer leading to about 6% of the magnetic field needed for field reversal. The apparatus is being improved for another series of experiments in which more electrons of higher energy will be injected into the E-layer. Whether a closed magnetic well[1] can be achieved remains to be demonstrated. After that real work on the main problem of plasma confinement and heating can begin.

---

1 A closed magnetic well is a configuration of closed magnetic field lines in which the magnetic field increases in every direction.

Figure 5-32. The Astron (side view).

**WORLD DISTRIBUTION OF EFFORT
ON CONTROLLED THERMONUCLEAR RESEARCH**

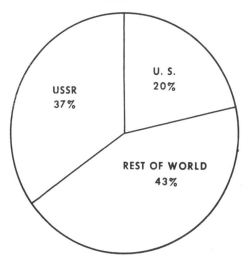

*Figure 5-33. World distribution of effort on controlled thermo-
nuclear research.*

REASONS FOR INTEREST
IN CONTROLLED FUSION RESEARCH

I PROGRAMMATIC JUSTIFICATION

ULTIMATE SOLUTION TO WORLD'S POWER REQUIREMENTS.   FEATURES:

1. Fuel is abundant, very inexpensive

2. Reaction products non-toxic, non-radioactive

3. No danger of run-away reaction

4. Long term possibility of direct conversion

II MORE BASIC INTERESTS

A. 99% OF UNIVERSE IS PLASMA

B. NEW AND UNEXPLORED SCIENCE

C. MANY SCIENTIFIC AND TECHNICAL POTENTIALITIES

Figure 5-34. Reasons for interest in controlled fusion research.

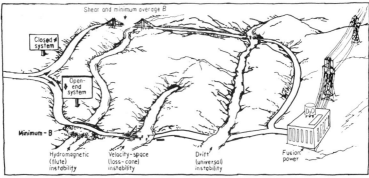

Figure 5-35. A few more rivers to cross to reach the Land of Fusion Power.

The Future

A considerable amount of attention is being given worldwide to the problems of controlling the thermonuclear reactions for the generation of electricity. *Figure 5.33* illustrates the world distribution of effort in this field. *Figure 5.34* lists some of the reasons for interest in controlled fusion research.

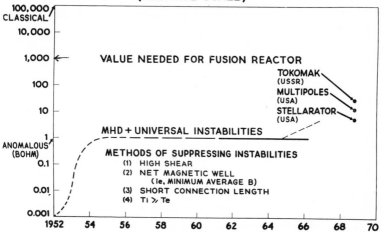

Figure 5-36. *Containment time in toroidal devices.*

CONSIDERATIONS OF FULL-SCALE FUSION REACTORS (I)

ASSUMPTIONS

1) MAJOR INSTABILITIES SUPPRESSED: PLASMA REASONABLY QUIESCENT
2) PRESENTLY-KNOWN SCALING LAWS HOLD

GENERAL FEATURES OF D-T REACTOR

1) REACTOR SURROUNDED BY SUITABLE BLANKET FOR ABSORBING NEUTRON ENERGY AND REGENERATING TRITIUM
2) CONFINING FIELD (FROM SUPERCONDUCTING COILS OUTSIDE BLANKET): ~80 KG
3) RADIUS OF VACUUM WALL: ~3M; OF MAGNET COILS: ~6M
4) POWER LEVELS: 5,000 MWe

	CLOSED SYSTEMS	OPEN SYSTEMS
5) T_i	~15 KeV	> 50 KeV
6) T_e	"	~ 30 KeV
7) FRACTIONAL BURNUP (f_b)	~6%	~8%
8) BETA (β)	~20%	~50%
9) REQUIRED CONFINEMENT TIMES	<1000 τ_{Bohm}	~10 $\tau_c^{90°}$
10) $Q \left\{ = \dfrac{\text{FUSION POWER}}{\text{INJECTION POWER}} \right\}$	–	~20

Figure 5-37. *General features of full-scale fusion reactors.*

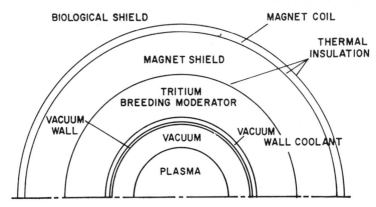

Figure 5-38. Configuration of a conceptual steady-state D-T fusion system.

We are making progress in our attempt to cross over into the Land of Fusion Power. There are only a few more rivers to cross. These allegorical rivers are shown on the map in *Figure 5.35*. As an example of our progress a bridge has been built over the flute instability river with open ended systems by using the principle of Minimum B. The same principle is assisting in bridging these instabilities in the closed system.

The containment time of plasmas in toroidal devices has long been limited at a value such as that labelled "Anomalous (BOHM)" on *Figure 5.36*. Recently containment times from 10 to 50 times this value have been achieved. As shown in *Figure 36* these containment times are becoming encouragingly closer to the value needed for a fusion reactor (roughly estimated as perhaps 1000 times the Anomalous BOHM values.)

It is, of course, too early to design a full scale fusion reactor. However, *Figure 5.37* lists some of the general features of a fusion reactor using the D-T reaction. *Figure 5.38* shows very crudely a configuration of a conceptual steady-state D-T Fusion System.

Cosmic Radiation

(Five Chapters)

by

C. B. A. McCUSKER

Professor C. B. A. McCusker,
Professor of High Energy Nuclear Physics, University of Sydney.

CHAPTER 1

Discovery

Cosmic radiation is the very energetic and penetrating radiation which falls on the earth from sources right outside the earth's atmosphere. In fact the nearest known source of some cosmic radiation is the sun and most of the radiation comes from very much farther away than that. The discovery of cosmic radiation came about through the investigation of a small and apparently trivial observation. This is not too unusual in science — the study of electricity, which has had enormous effects on all of our lives, started because a few people were puzzled by the way amber behaves when rubbed with a piece of silk. The relevant observation in the case of cosmic radiation was that ions are produced in a closed, gas-filled vessel at sea level. Probably the easiest way to show this is to take a gold leaf electroscope and charge it. The leaves diverge and stay that way for quite a long time (particularly on a dry day). But after a while they fall together because the charge on them has leaked away. If we improve the insulation and make sure that everything (including the gas) is dry we can increase the time before the charge disappears, but we can't completely stop it leaking away. By the beginning of this century, when people had learnt quite a lot about the conduction of electricity through gases, they realised that the leakage was due to the formation of *ions* in the gas, that is some of the atoms in the gas had electrons knocked off them so that in any given volume of gas there was a certain fraction of positive and negative charges. These could move to the gold leaves and neutralize the charge on them. At just about the same time Becquerel discovered radioactivity and people soon found out that the rays from radioactive substances ionised gases very easily. You can show that by bringing a radioactive source up to a gold leaf electroscope. The leaves collapse almost at once. So it was natural to attribute the ionisation that occurs to the small amount of radioactive materials in our surroundings. Then came the surprising observation. Sea water contains very little radioactive material — but the ionisation out at sea was not all that different to the ionisation over land. Now, to us, this seems a very conclusive observation. But, at the time (about 1910) it didn't attract much notice. This was partly because there were so many other interesting experiments being made in radioactivity — so many puzzles to unravel and results to think about. However, first Wulf tried a measurement of the ionisation at the top of the Eiffel Tower and then Göckel went up in a balloon. Both got the same result — the ionisation was not falling off as rapidly as it should do if it was all due to radioactive materials in the soil and rock. Göckel, in particular, came very close to discovering cosmic radiation but, unfortunately, he was using an ionisation chamber open to the atmosphere so, as he went up, the amount of gas in his apparatus decreased. The matter was settled by Hess in 1914 who flew to 5350 metres above sea level with properly sealed ionisation chambers. He found, to his great surprise, that the ionisation at 4000 m. was six times that at sea level and increasing rapidly with height. He drew the correct conclusion that there must be a very penetrating radiation incident on the earth's atmosphere from outer space.

Now at that time people knew four types of "radiation". There were X-rays which are photons of energy around ·01 to ·1 MeV. These were generated by allowing energetic electrons to hit a solid target and had been known for about 20 years. Then there were the "radiations from radioactive substances" called α, β and γ rays. α rays were helium nuclei — with energies around 5 MeV. β rays were just energetic electrons (typically around 1 MeV) and γ rays were, like X-rays, photons but much more energetic than the X-ray machines of those days could generate. Like the α particles their energies were around 5 MeV. Of all these types of radiation the γ rays were by far the most penetrating. α particles will penetrate a few centimetres of air and β particles a few 10's of centimetres but 10 MeV γ rays can be used to detect flaws in thick steel plate. Now the cosmic rays could penetrate our atmosphere — top to bottom. That is at least

1000 gms/cm^2, which is the same "mass thickness" as a 1 metre thick slab of lead. So people concluded that cosmic rays must be super powerful γ rays. They were wrong.

It took them a long time to find that out. For one thing, just after Hess' discovery, World War I broke out and stopped research into cosmic radiation (and into many other fundamental and interesting topics).

After the war it took some time for cosmic ray research to get going again and most of the early experiments emphasised the very penetrating character of the radiation. This seemed to make it more and more likely that it was a γ radiation. Then towards the end of the 1920's two very different experiments were carried out which completely changed the picture. The first was done by Clay—a Dutch physicist. He took an ionisation chamber by sea from Holland to the Dutch East Indies and measured the ionisation along the route. He found that it varied remarkably — that it was at least on the magnetic equator and increased as one went north or south of that line. Now γ rays are not deflected by a magnetic field but charged particles are. So straight away, here was a proof that a large part of the cosmic radiation must be due to charged particles. Now, charged particles are deflected one way if they are positive and the other way if they are negative, so by looking to see if more cosmic rays are coming from the East or the West at a given place one can tell the sign of the charge. When the experiment was tried a little later on it was found that most, if not all, the cosmic radiation must be positively charged. Since hydrogen is the commonest element in the universe — this made it likely that at least a fair amount of the beam was protons and since the range of protons in air is not very different to the range of a particles of the same energy it meant that the cosmic ray protons are much more energetic than any protons that had been encountered so far.

By the time all these consequences of Clay's simple but beautiful experiment had been worked out many of them had been confirmed by the other development. This was the introduction of new instruments and techniques. Up till 1928 the ionisation chamber (really just a development of the gold leaf electroscope) had been the only way of studying cosmic rays.

But in 1928 people started using Geiger counters and Wilson Cloud chambers. The Geiger counter is a device which detects ionising particles by greatly amplifying the tiny ionisation charge they produce. The resulting voltage pulse is quite large — 20 or 30 volts, if necessary, and also (for those days) quite fast, about 100 micro seconds. Now, a γ ray is not charged, so normally doesn't produce any ions in a Geiger counter. Once in about 1000 times it will knock a single electron out the counter wall and this can then discharge the counter. But protons and α particles are charged and they produce hundreds of ions when they go through a counter. So a proton is almost always detected Now a Geiger counter at sea level clicks away quite merrily. So either there are a reasonable number of charged particles around at sea level — or a thousand times as many high energy γ rays. There is a very easy way to decide this. Suppose we put one Geiger counter on top of another Geiger counter and arrange an electrical circuit so that it only gives an output when both are discharged simultaneously. If a charged particle goes through both — both will discharge. But if a γ ray goes' through No. 1 and is lucky enough to set it off it will only set off No. 2 once in a thousand times. So for a γ ray beam coincidence counts should be very rare. But in fact, they are not — so the beam must be mostly charged particles.

With Wilson Cloud Chambers it is even easier — a charged particle leaves a visible trail in a cloud chamber and in 1928 these were photographed by Skobeltzyn in Russia.

So by 1930, cosmic radiation was "on the map". People knew it existed; they knew it was made up of very energetic charged particles (some of which, at least could penetrate the whole of our thick atmosphere) and they knew that the charge was predominantly positive.

All this aroused a lot of interest and this interest lay in two main directions. Some people considered this enormously energetic radiation and wondered just what it was, and how energetic it was and where it came from and how it got accelerated. This meant they were really new sorts of astrophysicists. Other people looked at the great energies of the particles and compared them with the best energies that accelerating machines could produce

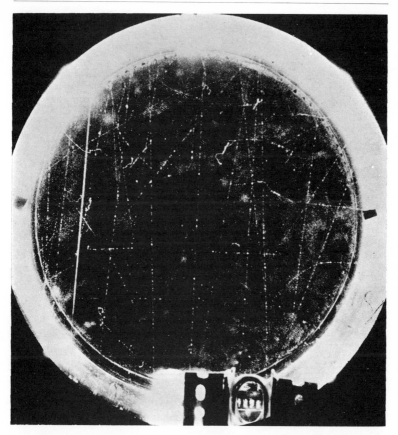

Figure 1-1. This is a Wilson cloud chamber picture of the cosmic radiation. Each of the tracks is made up of a string of drops of alcohol and water which are condensed out of the alcohol water vapour when it was suddenly cooled (by expansion) just after the shower of cosmic ray particles had gone through the chambers. The straight tracks are due to fast particles; the wiggly tracks are due to slow particles that scatter a lot more.

and realised that cosmic radiation far outstripped the machines. So they began to use cosmic radiation to study nuclear physics. These people became the first "particles physicists" as we would call them today. And, as we will see in the next few lectures, both sets of people made a host of interesting and fundamental discoveries.

CHAPTER 2

The Great Particle Hunt I

By 1930 then, cosmic radiation research was going strong. At that time people knew of two fundamental particles — the proton and the electron. Some of them suspected that there must also be a neutral particle of about the same mass as the proton and quite soon Chadwick, in Cambridge, discovered this particle. This made it possible to understand a great deal about the structure of atoms and nuclei. Atoms of ordinary hydrogen are made up of just one proton as the nucleus and just one electron revolving around it; one positive charge and one exactly equal negative charge giving a total charge of zero. But there is a fairly common heavy isotope of hydrogen whose atoms are almost exactly twice as heavy as those of ordinary hydrogen. Obviously this has a nucleus of one proton plus one neutron — total charge still just plus one — and one electron revolving round it. Since the chemical properties of an element are almost entirely determined by its electrons, heavy hydrogen behaves chemically just like ordinary hydrogen. But physical properties that depend on the mass of the molecules, like diffusivity for instance, are different for the two.

The next element is helium. Normal helium has an atomic weight of four and two electrons revolving round its nucleus. So obviously its nucleus is made up of two protons + two neutrons. And so we can go on up the periodic table — nicely explaining atomic weights and charges and isotopes. But, although one difficulty was thus settled, another one arose. The question was very simple: "What holds protons and neutrons together in a nucleus?" It can't be an electro-magnetic force because neutrons have no charge. The only other force then known was gravitation and this is much too weak. So there had to be another, third force. It had to be very strong because it takes much more effort to break up a nucleus than it takes to knock electrons out of atoms (millions of electron volts rather than tens of electron volts). On the other hand it had to have a much shorter range otherwise all the matter in the universe would be pulled together in one gigantic nucleus. A theory that incorporated these features was worked out by a young Japanese physicist named Yukawa. You know that the gravitational force between two masses varies inversely as the square of the distance between them. Or if you consider the *potential* rather than the force it varies inversely as the distance between them. Similarly for the electrostatic force between two charges. It varies as $\frac{1}{r^2}$, the potential varies as $\frac{1}{r}$. Now Yukawa wanted a potential that fell off more quickly than that so he put in an exponential term to this law and wrote the *nuclear* potential as

$$\Phi = C \cdot \frac{1}{r} \cdot e^{-\lambda r}$$

C and λ are constants. C just depends on what units we are working in. λ is more important — it fixes the *range* of the force. Suppose we measure r in units of 10^{-13} cms (often called a Fermi). Suppose we find $\lambda = 1$ in this system [so $\lambda = 10^{13}$ cm^{-1}] then Table I shows how $\frac{1}{r}$, e$^{-\lambda r}$ and their products vary with r. You see that for small values of r,

$\frac{1}{r}$ and $\frac{1}{r} e^{-r}$ are very similar but once r becomes bigger than unity, $\frac{1}{r} e^{-r}$ falls off very quickly. This, of course, is just the behaviour we want.

However, it has some interesting consequences. We know that the electro-magnetic field has a "particle" associated with it. This is the *photon*, the particle of light, which of course travels at the velocity of light. This in turn means that it has zero rest *mass*. Now we could write the electrical potential as

$$\Phi = C^1 \frac{1}{r} \cdot e^{-\lambda_e r}$$

providing that we put $\lambda_e = 0$.

r	$\cdot 1$	$\cdot 2$	$\cdot 5$	1	2	5
$\dfrac{1}{r}$	10	5	2	1	$\cdot 5$	$\cdot 2$
e^{-r}	$\cdot 905$	$\cdot 819$	$\cdot 607$	$\cdot 369$	$\cdot 135$	$\cdot 007$
$\dfrac{1}{r} e^{-r}$	$9\cdot 05$	$4\cdot 10$	$1\cdot 21$	$\cdot 37$	$\cdot 07$	$\cdot 001$

Table I. This shows the way in which a $\dfrac{1}{r}$ potential and a $\dfrac{1}{r} e^{-r}$ potential falls off as r increases.

This may sound a bit crazy but it does suggest that λ is associated with the rest mass of the particle belonging to a particular field. A little more detailed argument shows that this is correct and it follows that the particle associated with the nuclear field has a finite rest mass because λ is finite. It turns out that this mass is intermediate between that of the electron and the proton so the particle is called a *meson*. Yukawa also postulated, in order to account for the β decay of radioactive nuclei, that it should decay (when free) with a lifetime of $\backsim 10^{-8}$ seconds.

Now once this prediction had been made cosmic radiation was the obvious place to look for the meson. This was partly because no accelerator then existing had sufficient energy to make mesons [the rest mass was predicted to be 137MeV so at least that much energy was needed] but also because cosmic ray workers had already turned up a new particle. This was the positive electron (or positron). It too had been predicted, this time by P. A. M. Dirac, the famous British theoretical physicist. It was found by Anderson of Cal Tech using a combination of Geiger counters and a Wilson Cloud Chamber in a magnetic field. The cloud chamber was triggered when the Geiger counters discharged — an economical system that produces a high rate of useful pictures. The magnetic field is parallel to the axis of the cloud chamber. Any charged particle moving perpendicularly to this axis is deflected into a circular path. The direction of the deflection gives us the sign of the charge and the radius of curvature is a measure of the momentum of the particle. By combining this with the amount of ionisation per unit path length we can determine the mass of the particle. In this way Anderson and Neddermeyer were able to prove with just one photograph that positrons exist.

So when the meson was predicted cosmic ray physicists began to look for it. Once again it was Anderson who found it — or something so like it that for years people thought it was the real thing. What Anderson and his team turned up was a particle whose mass was about 200 times the electron mass, and which was radioactive with a lifetime of $\backsim 10^{-6}$ seconds. This was very like the postulated meson. But as people began to investigate it in more detail a queer thing came out. This particle which was supposed to hold the proton and neutron together in the nucleus and thus had to interact violently with them, this particle could pass through large quantities of matter without any sign of interacting other than the production of ionisation due to its electric charge.

At this point World War II broke out and most cosmic ray research came to an end.

When the war ended physicists came back to cosmic radiation with a lot of very much improved equipment — and with a great urge to solve this meson problem.

A lot of experiments were started. The one that paid off was beautifully simple. It was carried out by Lattes, Occhialini and Powell of the University of Bristol, England. During World War II, Ilfords had developed photographic emulsions as detectors of ionising particles. You may remember that it was by using photographic emulsion that Becquerel discovered radioactivity in the first place. But the old emulsions were very inefficient and would not produce well defined tracks of particles as would a Wilson Cloud Chamber. The developed emulsions were much better.

But, of course, the grains of silver bromide affected by the charged particle were very small (of the order of one micron or less) so the tracks had to be examined under high magnification. However, once this technique was developed the experiment was quite easy. A packet of photographic plates was left on a mountain top for a month or so (remember how the cosmic ray intensity increases as we go upwards) then taken down, processed and scanned under a microscope. Now the lower the momentum of a particle the more it is scattered from a straight line path by the nuclei it encounters. So scattering is a measure of momentum just as is curvature in a magnetic field. And the ionisation can be found by counting the number of developed silver grains per unit path length. So just as in the case of the magnetic cloud chamber an estimate can be made of the mass of the particle.

It is not a very accurate estimate but it could easily tell the difference between an electron, a meson and a proton (roughly 1 : 200 : 1800). The Bristol physicists found a peculiar situation. Some meson tracks came to an end and produced, at the end, a nuclear disintegration — with four or five particles flying outwards (from its appearance they called this sort of event a "star"). Some meson tracks stopped and apparently did nothing, and some stopped and produced a single new track of a particle also having mass about 200 times the electron mass and always of one particular length! This looks like a complicated set of observations to have to explain but the Bristolians managed to do it. First, they knew that their emulsions were too insensitive to detect lightly ionising particles like fast moving electrons and they knew that Anderson's meson decayed to electrons. So they said that their second group — the mesons that stopped and apparently did nothing were Anderson type mesons with a positive charge which came to rest, were repelled by any nuclei they got near, and so just sat there till they decayed with an electron and a couple of neutral particles all of which were missed by the emulsion. The first group, the mesons that produced the star, were genuine Yukawa type mesons with a negative charge which, once they had slowed down enough were attracted into the nearest positive nucleus, reacted violently with it and blew it up! The third type were also genuine Yukawa mesons but with a positive charge.

When they slowed down to rest (or more likely to "thermal" velocities) they couldn't get into any nuclei because of the electrostatic repulsion. So they wandered around slowly till they decayed, with the correct predicted lifetime of almost 10^{-8} seconds. But, and here was the surprise, they didn't decay to an electron but to an Anderson-type meson and one neutral particle. Because there was just one neutral particle and because energy and momentum have always to be conserved, this meant that the Anderson meson always came off with the same velocity and hence had the same range.

Well, nowadays, Yukawa-type mesons are called π-mesons or pions. Anderson-type mesons are μ-mesons or muons. There are three pions, π^+, π^- and π°, but only two muons, μ^+ and μ^-. All the pions interact furiously with protons and neutrons. Muons just ignore the strong nuclear force.

This work was published in 1947 and for a few months in that year everything looked fine. We knew that nuclei were made up of protons and neutrons — that atoms were made up of nuclei, plus enough electrons buzzing round them to balance the positive charge due to the protons in the nucleus. We knew that the protons and neutrons were held together by the pion "glue" and the pion had just the lifetime, mass and general properties predicted by Yukawa. Admittedly it decayed in a rather unexpected way — to another new particle, the muon. Or rather π^+ and π^- did this. π° decayed to two γ rays with a very short lifetime. But this wasn't a serious discrepancy. So nuclear physics looked nice and neat and orderly. But in the same year, Rochester and Butler in Manchester produced two cloud chamber pictures of two new and unexpected particles. When they were investigated they were found to have such peculiar properties that they were called the "strange" particles. And we are still trying to explain them.

CHAPTER 3

The Great Particle Hunt II

At the end of World War II, G. D. Rochester and C. C. Butler in Manchester, England, began an investigation of what were called "local penetrating showers" of the cosmic radiation. These showers were really high energy nuclear interactions — the initiating particle having an energy around 20 GeV (i.e., 2×10^{10}eV). They had been discovered by Wataghin — a Russian refugee working in Brazil. What was known about them was this — they were produced by one particle, sometimes charged, sometimes neutral; some of the secondary particles were very penetrating — easily passing through 20 cms of lead — some of the particles seen, however, were electrons and positrons. It was suspected that some of the secondaries were the Yukawa mesons that people were looking for. Remember, when Rochester started the experiment the pion had not been discovered. The idea of the experiment was to arrange Geiger counters in thick lead shields so that they would respond mainly to these rather rare "local penetrating showers" and to have a Wilson Cloud Chamber in a strong magnetic field right in the middle of this Geiger array so that some estimate of the mass of the various particles could be made. This straightforward part of the experiment went quite nicely — but in addition to this they got two pictures (in about three years) that could not be explained in terms of any known particles.

The first showed a pair of tracks shaped like an inverted V. The very small curvature of the tracks in the magnetic field showed that they were very energetic and the sign of the curvature showed that (if they were both coming from the point of the V) they had opposite charge. Of course the two tracks would be due to one particle if it had been coming upwards and collided with a nucleus at the tip of the V, being scattered through a very large angle as a result. But, in the first place it is unlikely that such a high energy particle would be coming upwards. Secondly, if it did collide with a nucleus then the nucleus must recoil — since momentum must be conserved and it's easy to work out the change in the particle's momentum. But recoiling nuclei of the necessary energy leave very thick prominent tracks in cloud chambers and there was no sign of any such track at the apex of the V. So they were left with only one explanation. The V must be due to two charged particles, one negative, one positive, resulting from the decay of a heavy neutral particle in flight. They could even estimate the mass of this particle and its lifetime. It had to be much more massive than the then known meson (the muon) and its lifetime was probably short, less than 10^{-8} seconds.

The other track was different. It looked like a very shallow V and they were able to show that it was probably due to the decay in flight of a massive charged particle.

Now, theoretically, this was all very unexpected. Not quite unexpected experimentally because a year or two earlier two Frenchmen, Leprince-Rinquet and l'Heritier, obtained a cloud chamber picture of a cosmic ray particle which indicated that it might have a mass about 1000 times that of an electron. Unfortunately their particle was not so obliging as to decay within the cloud chamber.

Now once we could be sure of the existence of these particles the sensible thing to do was to continue the experiment on a mountain top. These particles were obviously being produced by the cosmic radiation in high energy nucleon interactions. As one goes up, the cosmic radiation gets more and more intense and at 10,000 feet it is up by a factor of about thirty. But not every one was convinced by Rochester and Butler's pictures — and there are no 10,000 ft. high mountains in England, and Europe was still in rather a turmoil after the war. So the Manchester group was a bit slow in getting high up. But Anderson, still going strong in Cal Tech, was not. Since the war he and his group had continued their cosmic ray work with a beautiful experiment that showed that the muon decayed to at least three particles — one electron and two neutrinos in all probability. On hearing the news from Manchester they very quickly got a cloud chamber up to 11,000 feet and in 1950 published a paper giving 34 examples of the V particles.   Very soon after that mountain laboratories became crowded with cosmic ray physicists.

Meanwhile, however, the newer technique of photographic emulsions was producing results in the same field.   Following the discovery of the pion, nuclear physicists flocked to Bristol from all over the world.   They went to learn the new technique and to join in the many exciting experiments going on there.

In one of these experiments — again in a stack of nuclear emulsions that had been exposed to the cosmic radiation, they found a particle which came to rest and then produced three other tracks.   Now these could have come from a nuclear inter-action.   But one circumstance made this at once unlikely — all three secondary particles were in the one plane.   In a decay this is bound to happen, for the directions of any two of the particles define a plane and, obviously, their resultant momentum must be in this plane.   So to balance momentum the path of the third particle must be just in the opposite direction — and therefore, also in the plane.   But in a nuclear interaction one almost always has some neutral particles produced — neutrons or γ rays or neutrinos or neutral mesons of one sort or another.   So the momenta of the charged particles don't have to balance amongst themselves and they are hardly ever coplanar.

So the Bristol group investigated further and they were able to show that the original heavy meson was certainly moving in towards the star and two of the decay products were certainly pions and so on.

And in the end they were able to show that they had a heavy meson, mass $\backsim$ 970 times the electron mass, which probably decayed to 3 charged pions.

This really set things moving.   Quite unexpectedly, in two years since the pion was discovered we had one heavy neutral particle that decayed to two charged particles, one heavy charged particle that decayed to one charged particle and one or more neutral particles and one heavy charged particle that decayed to three charged particles.   There was an enormous outburst of research in cosmic radiation, partly done with cloud chambers, often in magnetic fields, generally on mountain tops and partly with nuclear emulsions, at first exposed on mountain tops and then later flown in balloons to ever greater heights.   These started at about 60,000 ft.

Figure 3-1. The start of the Sydney University 20 litre stack flight. The flight was organised by the Balloon Group at Holloman A.F.B., New Mexico. The balloon reached 126,000 feet and stayed there for 6½ hours. The stack and its parachute were cut loose by radio command when the balloon was about 300 miles from the launch point. The recovery crew reached the stack only 10 minutes after it landed.

and got higher and higher until nowadays balloon flights to over 140,000 ft. are fairly common. Also the balloons got bigger and bigger. Nowadays one can get 5,000,000 cubic foot balloons that are around 400 feet high when they are launched.

All this research produced some fantastic results. First, the Manchester group showed that one of the neutral particles was heavier than a proton — it was pretty obvious that it had to be because one of its decay particles was a proton! This started a new class of particles, all heavier than protons which were called *hyperons*. Nowadays it is a very populous class. This first of the hyperons is called the Λ° hyperon and its decay is written like a chemical equation

$$\Lambda^\circ \longrightarrow p^+ + \pi^-$$

Notice that electric charge balances on the two sides of the equation. Notice also that mass-energy also balances providing one counts in the kinetic energy of the proton and the pion. That is, we consider the situation in a frame of reference in which the Λ° is at rest before it decays. Then, when it decays the pion goes shooting off in one direction with a certain kinetic energy and the proton goes shooting off in the other direction. So we then have, using the famous Einstein law,

$$E = mc^2$$

(Mass of Λ°) $\times c^2 =$ (Mass of proton $+$ Mass of pion) $\times c^2$
$+$ K.E. of proton and pion.

Also people found that not all of the neutral V particles were Λ° hyperons. Some decayed to two pions and their mass was around 1000 electron mass, almost halfway between the pions (270 m_e) and the proton (1800 m_e).

The situation regarding the charged V particles was even more complicated. Some of them had masses around 1000 m_e — but it was hard to get good measurements and the error was quite big. This put them into the same mass range as the Bristol particle which decayed into three charged particles.

And then a charged meson was seen in a cloud chamber decaying in flight into three particles and an Irishman called O'Ceallaigh (pronounced O'Kelly) working in Bristol found a heavy meson in photographic emulsion which decayed into only one charged particle! In the five years from 1950 to 1955 things got more and more interesting and more and more confused. Of the heavy charged mesons that decayed into single charged particles (+ some neutrals, of course) some decayed into a pion, some into an electron, some into muons. But one set decayed to muons of a fixed range (which means that only one neutral particle is involved), others seemed to decay to a muon of variable energy. Along with all these decay modes there seemed to be a whole spectrum of masses varying from 500 to 1400 times the electron mass. Things were really very confused and the nuclear emulsion groups decided to make a great joint effort to straighten things out. Led by the Bristol group they flew a large emulsion stack (called the G stack) in the Mediterranean. Then they parcelled it out amongst various laboratories to make a detailed study, amongst other things, of this heavy charged meson. A conference was called at Pisa, in Italy, in 1955 to hear and discuss the results of this great collaboration.

Now while the cosmic ray physicists had been discovering and investigating all these new particles, the men who designed and built accelerators had been increasing the energy these could produce. In 1938 this limit was about 50 MeV, in 1947 the big accelerator in Berkeley reached 400 MeV. This is quite enough to produce pions in large numbers and the detailed investigation of pion properties soon passed into the hands of accelerator physicists. The reasons for this are simple. Once a machine has enough energy to produce a given particle it can make them in large numbers compared with cosmic radiation. Also they are produced when you want them, and where you want them — going in the desired direction with a given momentum and so on. The accelerators, in fact, have all the advantages except that their energy is limited. By the early 1950s the accelerator energy got up to 2000 MeV [= 2 GeV] and early in 1955 the Bevatron began working at Berkeley with an energy of 5 GeV. Very quickly, the Berkeley physicists got a *beam* of these heavy charged mesons and they started the Pisa conference with an account of the experiments they had done with this beam. It was obvious that for the study of all known particles cosmic radiation was outmoded.

It turned out indeed that there was only one heavy charged meson [that is, one and its anti particle, one with positive charge

Figure 3-2. The modern method of seeing hyperons. Instead of using cosmic radiation and a Wilson cloud chamber these physicists have used an accelerater and a bubble chamber. The interesting event is just to the right of centre in the top half of the picture. The initiating particle is coming vertically downward. It collides with a proton to produce a K meson and a hyperon. The hyperon is the short thick track going right. After a short time this hyperon decays to give a pion going left which in turn decays to a muon, and a neutral $V°$ hyperon which leaves no track but which in turn decays to the V shaped pair of tracks. Notice the curvature of the tracks in the magnetic field and the heavier ionisation of the slow tracks.

and the other with negative charge] and it had *six* different ways
of decaying.   This particle is now called the K-meson or kaon
and we can write

$$K^{\pm} \longrightarrow \pi^{\pm} + \pi^{\pm} + \pi^{\mp}$$
$$\longrightarrow \mu^{\pm} + \nu$$
$$\longrightarrow \pi^{\pm} + \pi^{\circ}$$
$$\longrightarrow e^{\pm} + \nu + \nu$$
$$\longrightarrow \mu^{\pm} + \pi^{\circ} + \nu$$
$$\longrightarrow \pi^{\pm} + 2\pi^{\circ}$$

Of course, this discovery didn't by any means solve everything.
In fact it raised one enormous difficulty.   It seemed impossible
to many theoreticians that the particle that decayed to the three
charged pions could be the same as that which decayed to two
pions.   This in turn led, within a year, to the famous discovery
of the Non-Conservation of Parity (or as Pauli said, the discovery
that God is a weak left-hander).

But the intervention of the accelerator did mean that cosmic
radiation was finished as a means of investigating fundamental
particles in the energy range below 30 GeV.   Many cosmic ray
physicists switched to accelerators — many went into the study
of the astrophysics and solar physics of cosmic radiation and a
few stayed on to look at the very high energies.

In the last two or three years there has been a slight drift
back to cosmic ray experiments on particles.   As a result of the
beautiful experiments with accelerators we now know of about
two hundred different particles (which used to be called elementary
or fundamental particles — but these seem the wrong words when
there are so many of them).   With so many to go at theoreticians
have tried to find some underlying order.   One of the ideas that
has been very successful is the idea that some sort of super-
fundamental particle exists which has been christened a 'quark'.
There are supposed to be three of these and they are christened
p, n & λ quarks and each has its anti particle.   It is possible to
make *all* the two hundred or so particles out of suitable combinations of these.   But to do so, in the simplest way, the quarks
have to have electric charges that are $\frac{1}{3}$ or $\frac{2}{3}$ of the electron charge!
This theory has been so successful at predicting the behaviour of

particles that many experiments with accelerators were set up to look for quarks. None of them have so far succeeded. Now this might mean that the accelerators just were not energetic enough — just as a 50 MeV cyclotron couldn't produce pions.

So people turned to cosmic radiation. But here again (up to December 1968) no one had been able to find a quark, with its queer charge, in cosmic radiation. But if they did, and if it happened only at high energies then particle physicists would have to come back to the cosmic ray studies they left in 1955.

CHAPTER 4

Nature and Origin

You remember that Clay showed, by taking an ionisation chamber across the Equator, that the cosmic ray intensity depended on magnetic latitude and hence that most of the primary radiation, at least, must be charged. And that a subsidiary experiment which found the difference in intensities coming from the East and the West at one place showed that most of the radiation must be positive. Since people like simple hypotheses, most cosmic ray physicists through the 1930s believed that cosmic radiation was just very high energy protons. In fact, although there are a lot of high energy protons in the cosmic radiation there are also many other nuclei. This was first shown by an experiment carried out by Martin Pomerantz shortly after World War II. He flew a Geiger counter telescope with 4 Geiger counters in it. Three were normal, high efficiency counters which were almost certain to respond when a singly charged fast particle went through them. But the fourth was deliberately made inefficient so that it only had a small chance of detecting protons. But ionisation rises as the square of the charge so this counter had a 50% chance of responding to a helium nucleus and better than a 96% chance of responding to lithium or anything heavier. By counting the rate of 3 folds to 4 folds he was able to show that there must be heavy particles in the cosmic ray beam.

Once photographic emulsions came into common use it was possible to go a lot further. By flying emulsions in balloons people soon found there are the nuclei of many different elements in the beam. Quite quickly most of the elements up to and including iron were noted. Hydrogen was the commonest, followed by helium. Just what the proportions are depends on how you measure it. For any one element the numbers of nuclei fall off rapidly with increasing energy. For instance protons of more than 10 GeV are about thirty times more common than protons of more than 100 GeV, so we can write

$$N(p, >E) = A \cdot \frac{1}{E^{1.5}}$$

where $N(p, >E)$ means the number of protons per unit time, per unit area, per unit solid angle with energy greater than E. This is called the integral, energy spectrum. This is all right for protons — but what about heavier nuclei? When we measure their energy do we take the *total* energy of all the protons and neutrons that make up the nucleus, or do we take the energy per nucleon or what?

It turns out that there are two schemes, one used at low energies (1 GeV to 100 GeV) and the other at high energies (> 10,000 GeV).

The low energy physicists generally compare the numbers of different types of nuclei at the same *magnetic rigidity*. This is because the earth's magnetic field acts as a giant separator and at any particular point it is the curvature of the path of the particle in the field which determines whether or not it can reach that point. Now *magnetic rigidity* is almost the same as *energy* per nucleon. This is because the curvature depends on the ratio of charge to mass and since most nuclei have about equal numbers of neutrons and protons this is the same for most of them. But ordinary hydrogen is different — its nucleus has one proton and no neutrons so its charge to mass ratio is twice as big as the rest. But apart from the protons which differ by this factor of two, magnetic rigidity and energy per nucleon are about the same.

At high energies we work at a given *total* energy because the total energy is comparatively easy to measure whilst trying to estimate what sort of a particle the primary was can be quite difficult. You will see that if we had a proton and an iron nucleus [Atomic Weight = 56] with the same *total* energy then the energy per nucleon of the iron nucleus would be 56 times smaller.

Now suppose we look at the low energy composition. At a given magnetic rigidity (corresponding to something like 4 GeV per nucleon) we find that about 80% of the particles are protons, about 20% are a helium nuclei (or *α* particles as they are still called) .and a couple of a percent are heavier particles. Now the *total* energy of these heavier particles will of course be much greater than the total energy of the protons so if we had measured at a given total energy the proportions of protons to α's to heavies would have been much more nearly equal. If we go on to look at the distribution amongst the heavy particles we find some very interesting things. This is shown in the Table. Here we are comparing the abundance of the elements heavier than helium in the cosmic radiation with their abundance in the universe at large. This is got from studying the earth, meteorites, the sun, other stars and clouds of gas and dust floating between the stars. You can see at once that the abundances in cosmic radiation are quite different. For instance the universe as a whole has very little lithium, beryllium and boron but cosmic radiation has quite a lot. Also cosmic radiation has much more iron, compared to hydrogen than the universe as a whole.

Element	Atomic Number	Abundance in Cosmic Rays	Universal Abundance
Lithium	3	3·9%	3×10^{-6}
Beryllium	4	1·7	5×10^{-5}
Boron	5	11·6	4×10^{-4}
Carbon	6	26·0	10
Nitrogen	7	12·4	16
Oxygen	8	17·9	39
Fluorine	9	2·6	$1·6 \times 10^{-2}$
Heavier than Fluorine	>9	23·9	35

Table I. The percentage abundances of the elements heavier than helium in cosmic radiation and in the universe generally.

All this means that cosmic rays must have originated in some place that is not typical of the universe as a whole — some place where there were a lot more heavy elements. Then after acceleration the radiation must have passed through about $3g/cm^2$ of matter before reaching the top of our atmosphere. The effect of this passage was to break some of the heavy nuclei, like iron, into smaller bits giving us the lithium, beryllium and boron in the primary beam.

The next thing we have to consider is the amount of energy in cosmic radiation. Cosmic ray particles aren't all that frequent, about one per square centimetre per second, at the top of the atmosphere. This is very few compared with the number of photons of starlight — but the energy of each cosmic ray particle is enormous compared with the energy of a starlight photon. So it turns out that there is a lot of energy locked up in cosmic radiation in our galaxy.

Now cosmic rays can, of course, escape from the galaxy (quite easily, some people think) so if they are produced here there must be some major energy sources in our galaxy making them. We know that our sun produces occasional bursts of low energy cosmic rays (up to about 50 GeV) but if we multiplied the sun's output of cosmic radiation by the number of stars in our galaxy we would be short by many factors of ten. We need something much more violent and drastic than our quiet well behaved sun. The obvious candidates are the supernovae.

A supernova is what happens when a sufficiently large star runs out of nuclear fuel. In any star there is normally a balance between the outward pressure of light and heat produced by the nuclear reactions going on in its exceedingly hot core and the gravitational in-pull of its great mass. When the star runs out of nuclear fuel one of these forces suddenly stops: the star collapses inwards in a matter of *seconds* and there is an enormous explosion. For a short time the light from the one star exceeds that of all the other ten thousand million stars in the galaxy. Energy is given off in many ways — huge amounts of matter are blown off into space — super shock waves tear through this expanding envelope at nearly the speed of light and the core of the star collapses inwards till the matter is as dense as that in an atomic nucleus.

This density is about 7×10^{15} gms/cc or 30,000,000,000 tons per cubic inch. This core 'bounces' a few times but this radial motion is highly damped and it soon settles down as a super atomic nucleus weighing about as much as our sun but only 7 kilometres across. But, of course, angular momentum must be conserved and just as a skater spins faster when she pulls her arms inwards so this core is rotating very much more quickly than the original star. Instead of turning once in 27 days as our sun does it will be whipping round at perhaps 100 times per second to begin with. Also, any magnetic field lines that were in the original mass are compressed into the tiny little core so the magnetic fields are colossal — perhaps 10^{13} oersted. This is what is called a neutron star because 90% or so of its matter is made up of neutrons; and it is these neutron stars which, we believe, give out the regular radio signal that we call *pulsars*.

Now the colossal magnetic field of the star will bind charged particles into a sort of super Van Allen belt around the star. These belts will be rotating with the star — as far as possible, that is, because you don't have to go very far from an object rotating 100 times per second to reach the speed of light, if you rotate with it. You have in fact the simple equation

$$3 \times 10^{10} = 2\pi \cdot r_c \cdot 100$$

which gives r_c, the radius at which the speed of light would be reached for a rigid rotator as about 500 kilometres. Now, of course, matter cannot reach the speed of light so what happens is that these outer particles lag behind and act as a sort of brake. Energy is lost, various sorts of radiation are produced, particles are accelerated up to cosmic ray energies. The neutron star gradually slows down. All this was predicted by Thomas Gold — before the slowing down was actually observed experimentally. This observation confirms that the pulsars are losing their rotational energy and there is a tremendous amount of it to lose — about 10^{52} ergs per pulsar—and the supernovae which produce the pulsars occur about once every hundred years in our galaxy. So here we have a quite sufficient source of energy to produce cosmic radiation and it comes in a very handy form — a very dense, small, rapidly rotating object with enormous built-in magnetic fields.

We are pretty sure, then, that a lot of the cosmic radiation must be produced by these fierce little objects. But does it all come from them? To some extent we'll save this question up for the next lecture but at least we can sketch in some of the things we have to consider.

For one thing, what sort of energy limit do we have to go up to? So far the highest energy we have talked about was around 1000 GeV. But we know that cosmic rays come with much higher

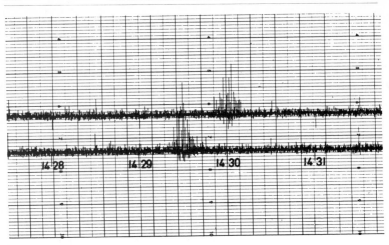

Figure 4-1. A pen recording of a radio signal from a pulsar made at the Sydney University Mills Cross telescope at Molonglo Observatory. This was the fifth pulsar discovered by Professor Mills, Dr. Large and their co-workers. By February, 1969, they had discovered another dozen. The upper and lower traces show two passes of the pulsar through the antenna beam.

energies. Up to about 100,000 GeV we can still see their individual tracks in photographic emulsion and know that some are still protons, some α particles and some heavier nuclei. And studies of the great showers of particles these super energetic rays make in our atmosphere enables us to go a lot further. We think that one shower of energy about 10^{11} GeV has been detected. Can pulsars make such high energies — as yet we don't know for certain but I don't think they can. I think there is some indication that they normally get up to about 2×10^6 GeV per nucleon.

Since the matter they are accelerating is taken from the surface of a neutron star this could quite easily mean 5×10^8 GeV for a uranium nucleus. Beyond that we might have to look at even wilder and more energetic places for our cosmic radiation.

CHAPTER 5

Very High Energies

The higher the energy we want, the fewer cosmic rays there are. If we consider energies greater than say 1000 GeV [the world's largest accelerator produces protons of 70 GeV] then we can get a reasonable number in a fairly big emulsion stack in one balloon flight. For instance, the Sydney 20 litre stack was flown in 1961 for $6\frac{1}{2}$ hours at 126,000' over New Mexico. In that stack we found 52 protons, 18 α particles and 42 heavier nuclei each having a total energy greater than 1000 GeV. If we go to energies greater than 100,000 GeV only about six have been seen in all the stacks ever flown (some as big as 80 litres). For energies greater than 500,000 GeV we have to use other methods of detection. Fortunately this is easy.

When a high energy particle hits the atmosphere it soon reacts with some nucleus in the atmosphere. Out of this collision come lots of mesons of all sorts, some hyperons and so on. Pions are produced copiously. Now some of these particles go on — still with plenty of energy to interact again. So we get a cascade process down through the atmosphere. But some of the charged pions decay before they can interact like

$$\pi \pm \longrightarrow \mu \pm + \nu$$

The muons don't have a nuclear interaction so they continue on as penetrating muons till they decay. The neutral pions decay in a different way and with a much shorter life time so they nearly always decay rather than interact. Their decay scheme is

$$\pi^\circ \longrightarrow 2\gamma$$

that is they produce two energetic γ rays. These γ rays start a cascade of their own, called the *electro-magnetic* cascade. The two basic processes are

$$\gamma \longrightarrow e^+ + e^- ,$$ in the electrical field of a nucleus close to the γ path and

$$e\pm \longrightarrow \gamma + e\pm$$ when the electron is deflected in the

field of a nucleus. The first process is called pair production and you notice that it is a conversion of *energy* into *mass* + energy. We have a particle (e⁻) and its anti particle (e⁺) produced. Charge, of course, is conserved. The second process has a German name, bremsstrahlung — meaning 'braking radiation' because it is an electro magnetic radiation produced by the deceleration of the electron. These processes can keep up a cascade process as long as the particles have enough energy and we get, for instance

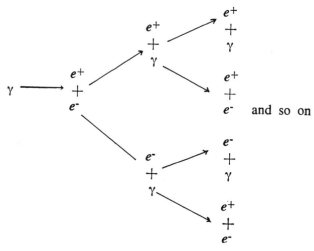

and so on

The cascading continues till the individual particle energy is somewhere around the *critical* energy. This critical energy varies from material to material. It is about 1 MeV for lead and 100 MeV for air. Below the critical energy the particles rapidly lose their energy by other processes (like ionisation) and the cascading stops. So now we have a picture of this whole process. The 'back-bone' is the nucleon cascade. This keeps on feeding energy into muons and into the electro-magnetic cascade. All the particles scatter somewhat from the direction of the original particle. The lighter they are and the less energetic they are, the more they scatter. Really high energy nucleons and pions are always within a few tens of metres from the axis. The electrons scatter out for a few hundreds of metres and muons spread so they are contained in a circle whose radius might be two or three kilometres at sea level.

The whole phenomenon is called an *extensive air shower* because it is a shower of particles, it happens in the atmosphere and it is quite extensive. One of the world's records we hold in Sydney is for the biggest distance between two stations hit by one air shower — 8 kilometres in shower No. 1622! These showers were discovered just before World War II by Pierre Auger. He found that if he had three trays of Geiger counters sitting side by side in his laboratory they sometimes went off together — obviously hit by a shower of particles. But equally obviously, the shower could have come from the laboratory roof. So he put one of the trays outside and some distance away and found that he still got some coincidences between trays — not as many as before, but still, some. When he increased the separation the coincidence rate dropped but did not disappear. Even when the trays were 50 metres apart he still got coincidences. This was a big distance for those days and he worked out roughly that the primary particle must have had an energy of about 100,000 GeV. No one had thought that cosmic ray particles of this energy existed

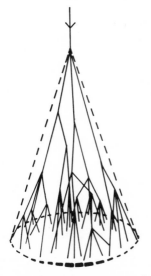

Figure 5-1. A schematic diagram of an extensive air shower. The particle coming in to the top of our atmosphere is shown by an arrow. The shower reaches sea level as a disc of particles whose radius is about 2 kilometres.

and it raised a lot of problems about where they came from and how they were accelerated. Also it promised a whole new field of very high energy nuclear physics for people to investigate.

After World War II many experiments were carried out on these large showers. The particles that Auger had detected were the electrons (and as a result it was at first thought that these were purely electro-magnetic showers perhaps produced by very high energy γ rays). But then, first the muons were detected and then the nucleons. To begin with Geiger counters were the main instruments used to detect them — with Wilson Cloud Chambers to examine the tracks.

But Geiger counters are slow old things. They take about a millionth of a second to begin to respond and about a hundred times that long to complete their pulse. People began to use scintillation counters instead. These consist of slabs of plastic, or dishes filled with special liquid, which give off a flash of light when a fast, charged particle goes through them. The light is picked up, turned into an electrical pulse and greatly amplified by a device called a photomultiplier. These scintillation counters respond in $\frac{1}{100}$ of a microsecond or less. Using these it was found that most of the charged particles in an air shower were contained in a disc only one or two metres thick (and radius up to 2 kilometres) moving perpendicular to the direction of the original particle with just about the speed of light. Once this fact is known we can use it and the fast response of the scintillators to get the direction of the showers just by timing the arrival of the shower front at different stations spread out on the ground.

The scintillators have another useful property. If two particles go through together we get twice as much light out, and so on. So we can use the *pulse height* from the photomultiplier to find out how many particles go through each station. And since the particles always cluster more thickly at the centre we can find out from the pulse heights and the direction just where the centre was. Finally, by sampling at a number of stations we can estimate the total number of particles in the shower and from this get the energy of the primary particle — all this from a few tiny, brief flashes of light!

Figure 5-2. One of the liquid scintillator tanks used in the Sydney University grand air shower experiment. The liquid, 10 cms deep, covers the bottom of the tank. The photomultiplier is placed where the stem meets the base. In the experiment the tanks are buried in the ground and detect muons. In all, 116 of these tanks are spread over 100 square kilometres in the Pilliga State Forest. This tank was being tested on the Sydney campus.

Now, while all this had been going on, the detecting arrays had been getting bigger — and in step with this, the largest showers seen kept increasing. The cosmic ray groups at M.I.T. built an array in New Mexico which had its scintillators each a kilometre apart — they covered an area of 8 square kilometres. And in four years running they got one shower with an estimated 5×10^{10} charged particles in it which means of primary energy quite close to 10^{11} GeV!

This at once raises the questions — are there particles even more energetic than 10^{11} GeV and, if there are, is it worth studying them? The first question can be answered if we can build a big enough array, say ten times bigger [80 km²] than the M.I.T. array at Volcano Ranch. To the second question, most cosmic ray physicists would answer "Yes" just on the grounds that anything that is an extreme is worth studying. But to make matters more interesting a prediction was made in 1966 that the cosmic ray spectrum would cut off — that 10^{11} GeV was the limit.

This prediction was made for the following reasons. For the last twenty years or so there has been two main contending theories about our universe. One theory said that the universe was infinite in space and time and was always much as it is now. This is called the Steady State theory. The other said that our universe was finite and (at least in its present state) began about 15,000,000,000 years ago with the colossal explosion of a gigantic, super hot fire ball. Somewhat irreverently, this was called the Big Bang theory. Now if the Big Bang happened then the fire ball would expand very rapidly and cool down very rapidly. The photons present would constitute what is called a Black Body radiation and as the universe expanded this radiation too, would cool down. At our present stage it should have cooled down a lot and be a universal Black Body radiation at a temperature of only $3°K$; or thereabouts. This should be all over the universe and at any point in it should be coming equally from all directions. Now we know just what energy and distribution of energies the photons in a Black Body have at any particular temperature.

For $3°K$ they should be predominantly in the radio microwave radiation (wavelengths between 10 & 0·1 cms.) and there should be about 600 per cc. Now we come to the connection with cosmic rays. Kenneth Greisen of Cornell pointed out that if we were sitting on a 10^{11} GeV proton moving through this gas then they wouldn't look at all like microwaves. Because the proton was moving so quickly through the gas they would look like very hard γ rays. And at that particular energy, close to 10^{11} GeV we would get the reaction

$$p + \gamma \longrightarrow N^* \ (1238)$$
$$\searrow \quad p + \pi$$

The proton would be turned into one of the short lived 'resonances' which in turn would quickly decay to proton and pion and, on average the pion would take 20% of the energy. So every time a proton got up to 10^{11} GeV it would promptly (i.e. within about 10^7 years) make a collision that put it 20% down the energy scale.

This looked to be a prediction really worth testing because in 1965 Penzias and Wilson at the Bell Telephone Laboratories had found what seemed to be the Black Body radiation.

Fortunately, by that time, we were already well on the way to having a big array. At the moment (Feb. 1969) we have 40 square kilometres operating in the Pilliga State Forest south west of Narrabri and by the end of 1969 we should have 100 km^2 going. It is the biggest cosmic ray detector yet built, by far, and with it we should be able to see if the spectrum cuts off as predicted, and to find out a good deal else besides.

Particle Physics

(Thirteen Chapters)

by

PROFESSOR W. K. H. PANOFSKY

and

PROFESSOR R. H. DALITZ

Professor W. K. H. Panofsky,
Director, Stanford Linear Acceler-
ator Centre, Stanford University,
California.

Professor R. H. Dalitz,
Department of Theoretical Physics,
Oxford University, Oxford,
England.

CHAPTER 1
(W. K. H. Panofsky)

The Purpose and History of Elementary Particle Physics

"Elementary Particle Physics" is the science of man's quest for an understanding of the ultimate nature of matter. This search is ages old, and today's "high-energy physics" is the culmination of twenty-five centuries of search. One of the purposes of this first talk is to explain why the drive toward higher energies in modern particle accelerators and the ancient search for the fundamental building blocks of matter are one and the same.

The earliest record of the question: "If you continue to divide matter into smaller and smaller units, when will it cease to be the same kind of material?" stems from Thales of Miletus (of about 580 B.C.). (Miletus is on the western Aegean coast of what is now Turkey.). He raised the question of whether the "indivisible" (ατομοσ — from which our word "atom" is derived) exists. Leucippus (460 B.C.) presented a much more vivid and modern picture — he considered the "indivisible units" to be "particles in a void" (then thought to be air). Leucippus also proclaimed an idea which has since had a strikingly modern rebirth: "Nothing happens without a cause, but everything happens with a cause and of necessity". In other words, it is just as important to know why something does *not* happen as it is to understand why something *does* occur. We will meet the importance of this thought much later when we talk about the "selection rules" which control the permitted transformations among the particles of modern physics.

Democritus (born 460 B.C.) put these concepts into even more concrete terms: There are many kinds of particles; they differ in size and substance; they are never annihilated; they move about "in the void" and *there is no other reality*. These ideas were summarised in the famous poem "de rerum natura" (of the nature of things) by the Roman Lucretius (Titus Lucretius Caro, 98-55 B.C.).

Although some of these ancient ideas we have been discussing were remarkable insights, we should not overestimate their true significance. For one thing, the old concepts that have retained

313

their validity until today were only a few of many philosophic speculations — most of which fell very wide of the mark. In addition, the good ideas were still only ideas: remarkable guesses about nature, but not backed up or tested by detailed experimental study. Perhaps the most important fact about the early Greek ideas was not how right these answers were, but rather that for the first time in recorded history, men had begun to ask themselves the deep questions about the world they lived in.

After these early beginnings, the rational search for "the nature of things" came to an almost complete stop for about 16 centuries, until Galileo (1564-1642) reaffirmed the belief that atoms moving in a void (not air by now!) determined the nature of matter. We then rapidly enter modern history — modern in the sense that ideas and experiment combine, and in the sense that the discoveries of the elements, of molecules, of atoms, and finally of the atomic nucleus have retained their validity until today. I will not dwell on the history of these discoveries which started with Boyle's (1627-1691) definition of an "element", and culminated with Lord Rutherford's demonstration in 1911 that the atom consists of a positively charged nucleus surrounded by negative electrons, with the nucleus being much smaller and containing most of the atom's mass. All of this material is the substance of the modern chemistry and atomic theory that you have already studied.

Lord Rutherford used high speed α particles (helium nuclei, made up of 2 neutrons and 2 protons) as projectiles to study atomic structure. These particles are emitted from natural radioactive substances and move at a velocity of about 1/50th of the speed of light. Their momentum is sufficient to penetrate the electron shell of an atom, but they are deflected without penetration by the electrostatic repulsion of the nucleus (*Figure 1-1*). To investigate the structure of the nucleus it would appear that we need enough kinetic energy E to get inside the nucleus of radius R, i.e.,

$$E > \frac{Z_1 Z_2 e^2}{R}$$

(1.1)*

where e is the electronic charge, Z_1 is the atomic number of the bombarding nuclei, and Z_2 is that of the target nucleus to be

---

* Equation (1.1) is in c.g.s.-e.s.u. units in which e is measured in statcoulomb which are 3×10^9 times smaller than the coulomb. In statcoulombs the electrical charge is 10^{-10}.

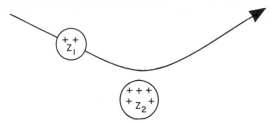

*Figure 1-1. Scattering of an α-particle (He nucleus); charge =
2e(Z_1 = 2) from a nucleus of charge $Z_2 e$.*

explored. This is one of the reasons why the study of nuclear
structure *requires high energy*; arithmetic substitution indicates
that for a proton ($Z_1 = 1$) to penetrate into a uranium ($Z_2 = 92$)
nucleus ($R \cong 10^{-12}$ cm) would require about 25 MeV, where
MeV stands for "million electron volts". This term means the
energy a particle carrying one electronic charge would have
acquired had it been accelerated by a potential drop of one
million volts.

We have thus identified one reason why the study of atomic
nuclei requires high particle energies: The bombarding particle
must penetrate the electrostatic repulsive barrier around the
nucleus. However, this is not the only, nor even the most important, reason. The primary reason for requiring high energies
is related to the dual nature of all the particles of nature: On the
one hand, particles have a mass m and a velocity v with which
they move in space. On the other hand, the probability of finding
a particle is controlled by a group of waves of wavelength λ given by

$$\lambda = 2\pi \, \hbar/mv \qquad (1.2)$$

where $\hbar$ is Planck's constant*

$$\hbar = 1 \cdot 054 \times 10^{-27} \text{ erg sec} = 1 \cdot 97 \times 10^{-11} \text{ cm MeV/c}.$$

This group of waves (*Figure 1-2*) moves with velocity v but the
particle cannot be more precisely localised than within one wavelength λ. Therefore, if we wish to study the details of nuclear
dimensions ($\sim 10^{-13}$ cm, or one Fermi), or even of sub-nuclear
matter, we require correspondingly large particle momenta. If the
wavelength associated with the incident particle is comparable to
the size of the object to be studied, the "wave packet" which

---

* Actually the constant first introduced by Planck, denoted by h, is equal to
$2\pi\hbar$. The constant $\hbar$ is the one which enters most of our discussion, however. It is the fundamental unit of angular momentum.

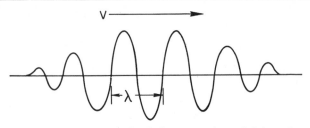

Figure 1-2.   A "wave packet" which governs the probability of finding a particle moving at velocity v.

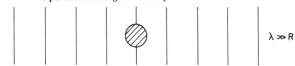

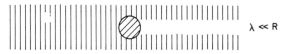

Figure 1-3.   Schematic representation of a wave impinging on an obstacle.   The two cases of wavelength larger or smaller than the dimensions of the obstacle are shown.

controls the probability of the particle being deflected into a specific direction will "diffract" around the object and little detail will be revealed.   This picture for the two cases

$$\lambda \gg R \qquad\qquad \lambda \ll R \qquad\qquad (1.3)$$

is shown in *Figure 1-3*.   This situation was put into more quantitative terms by Heisenberg in 1926 in terms of the "uncertainty principle", which we can write as

$$\triangle p \, \triangle x \approx \hbar. \qquad\qquad (1.4)$$

This equation applies to the situation of a "scattering experiment" (*Figure 1-4*).   Here $\vec{p_1}$ is the incident momentum of the particle, $\vec{p_2}$ is the final particle momentum, and $\vec{\triangle p} = \vec{p_2} - \vec{p_1}$ is the "momentum transfer" or "impact" to the unknown target object under study.   The quantity $\triangle p$ is just the magnitude of $\vec{\triangle p}$ and $\triangle x$ is the uncertainty in position or of geometrical detail to which the behaviour (such as the angular distribution of the scattered particle) of the process will reveal the position or geometrical structure of the target.

Figure 1-4.   *Change of momentum of a bombarding particle in a scattering experiment.*

We therefore conclude that the information about smaller and smaller distances will require particle beams of higher and higher energy. This situation is illustrated in *Figure 1-5*. We need light (energy ~1 *eV*) to see microbes (10^{-3} cm); we need electron beams of 1 *GeV* (1 *GeV* = 10^9 *eV*) to see some details of the nucleus of the hydrogen atom or the proton (10^{-13} cm); and we need even higher energy accelerators to get intricate, detailed information on the structure of the particles within the nucleus.

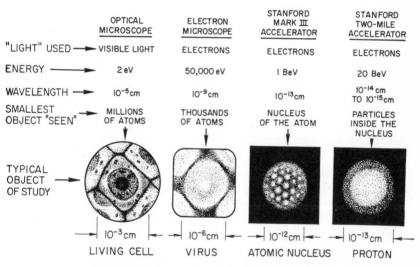

Figure 1-5.   *Comparison of four "microscopes".*

So far we have identified two reasons why the particle physicist needs high-energy particles: (1) to overcome the repulsive electric force between particles, and (2) to obtain enough "resolving power" to be sensitive to structure within the particles. Now let me add a third: (3) to have enough energy to "create" new, massive particles not ordinarily existing in nature.

Einstein postulated that mass and energy are equivalent, i.e., both are different aspects of the same thing. If a particle has a "rest mass" m_0 then its total energy E is given by

$$E = m_0c^2 + \text{kinetic energy} \qquad (1.5)$$

where $c = 2 \cdot 998 \times 10^{10}$ cm/sec is the velocity of light in a vacuum. To "create" a new particle we thus need an energy $E = m_0c^2$ plus whatever energy is needed to supply the kinetic energy of the particles generated in the reaction. Let us recall here the "rest energy" (i.e., m_0c^2) values of some of the particles measured in electron volts: m_0c^2 for the electron is $0 \cdot 511$ MeV or about one-half MeV (million electron volts); that of the proton is $938 \cdot 3$ MeV, or slightly less than 1 GeV (billion electron volts in the U.S. and in the Soviet Union, and Giga electron volts elsewhere). The rest mass of the first discovered "unstable" particle, the μ-meson or muon, is $105 \cdot 66$ MeV.

One might guess simple-mindedly that an accelerator of minimum energy of $105 \cdot 66$ MeV would be required to create a muon. This turns out to be wrong for two reasons: The first is that it is impossible to create new particles without at the same time giving some kinetic energy to the new particles and to whatever other particles happen to be around. The second reason is more profound: When "creating" new particles one has to "conserve" more quantities than just energy. You all know that electric charge is one such quantity. It can neither be created nor destroyed. It turns out, and this will be discussed in greater detail in subsequent lectures, that each of the particles of physics carry distinguishing labels (or "quantum numbers" as they are called in more highbrow language), and each of these quantum-number labels has to balance out in any transformation among particle systems. Let me remind you that those of you who study chemistry have already met with this concept. You have learned that atoms have a "valence", and that when you write a chemical equation you have to balance the valences. For instance, hydrogen has valence $+ 1$ and oxygen has valence $— 2$; therefore the stable compound of hydrogen and oxygen, i.e., water, has the formula H_2O.

Applied to the example at hand it turns out that muons can only be created in pairs. For each positive muon a negative muon has to be created, and that "costs" extra energy. Therefore the "threshold energy" needed to create a new particle is controlled not only by its rest mass, but also by the conservation laws which control the types of particles which must be produced together.

The accelerators of most interest in modern high-energy physics accelerate either electrons or protons, and the target nucleus is generally a proton, or a neutron bound in a nucleus. If we let m_0 be the rest mass of the incident particle of kinetic energy T, and M_0 be that of the target nucleon (proton or neutron, assumed at rest), then it can be shown that the energy U which is "available" for creating new particles is given by:

$$U = \sqrt{2M_0c^2T + (m_0 + M_0)^2c^4} - (m_0 + M_0)c^2 \qquad (1.6)$$

This equation is derived by balancing the momentum and energy of the incident and target particles, but using expressions derived from the theory of relativity for these mechanical quantities applicable to particles moving at very high speed.

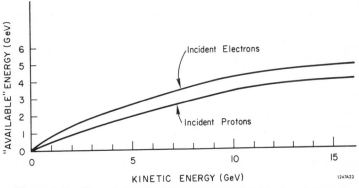

Figure 1-6.   Plot of "available" energy vs. the kinetic energy of electrons or protons bombarding a hydrogen nucleus (proton).

Figure 1-6 shows a graph of this equation plotted both for incident electrons and protons.   For very high energies, where $T \gg m_0c^2$ and $T \gg M_0c^2$, this equation becomes simply

$$U = \sqrt{2M_0c^2T} \qquad (1.7)$$

which shows that the available energy increases only with the square root of the bombarding energy.

We can easily visualise the physical situation (*Figure 1-7*). Before the collision the incident particle of rest mass m_0 approaches

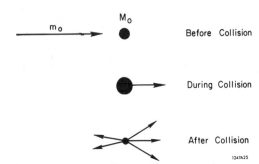

Figure 1-7. Sequence of collision phenomena where a particle of rest mass m_0 hits a target at rest of mass M_0, followed by disintegration into five particles.

the target particle of mass M_0. During the collision the particles "stick" together, but the centre-of-mass of the whole system moves ahead. The "available" energy U is the total energy of the system other than the kinetic energy with which the entire complex moves. The energy U is therefore often called the "energy in the centre-of-mass" or c.m. energy. Clearly the energy associated with the motion of the centre-of-mass is not of interest as far as the reaction itself is concerned, since it depends on the relative motion of an observer to the whole system. The energy U, on the other hand, is a fundamental quantity representing the energy excess that is associated with the compound system; it is a "relativistic invariant" in more highbrow language.

If U is large enough to create new particles, then the glob in *Figure 1-7* will come apart, and the final state will be a group of particles which on the average have forward momentum.

Let us be quantitative: To create a pair of muons in an electron-proton collision we need a bombarding energy $T = 236$ *MeV*, which is only 24 *MeV* more than the combined rest energy 212 *MeV* of the two muons which have to be formed. In contrast, consider the energy required in a proton-proton collision to create a pair of a proton and an antiproton, each of rest energy 938 *MeV*. We take

$$m_0 = M_0 \text{ and } U = 2M_0c^2$$

and get

$$T = 6M_0c^2 \text{ or } 5\cdot63 \text{ } GeV!$$

It is clear that, the higher the energy of the processes we are trying to study, the less efficient is the use of the bombarding energy.

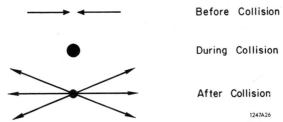

Before Collision

During Collision

After Collision

1247A26

Figure 1-8.   Sequence of collision phenomena for colliding beam particles of equal mass meeting head on.

Is there a way out?   Yes, there is, but it too has its problems. We can build machines called "colliding beam" accelerators in which the particles of *two* beams make a head-on collision (*Figure 1-8*).   In this case it is easy to see that *all* of the energy of the two incident beams is "available" for the collision, because the combined centre-of-mass is at rest.

The problem here is that man-made beams are not very dense — at best they carry about 10^{12} or so particles per cm^3.   In contrast, the density of condensed matter is much larger, e.g., liquid hydrogen (density ·07) contains 4×10^{22} atoms of hydrogen (and therefore protons) per cm^3.   Quite a difference!   Therefore, with colliding beams we can get much higher usable energies, but the number of interesting events we can study is much smaller.

Let us recapitulate here: The elementary-particle physics of today is the current frontier of the twenty-five-centuries-old quest to understand the ultimate constitution of matter.   But to pursue these studies experimentally we need accelerators of high energies — in the multi-GeV range — in order to

(a) bring inter-acting particles close together against their repulsion,

(b) be able to examine details of the structure of these particles at the small dimensions (less than 10^{-13} cm) involved, and

(c) create new, unstable particles, not occurring in nature.

In further discussions we will examine how these extremely high particle energies are obtained in practice.

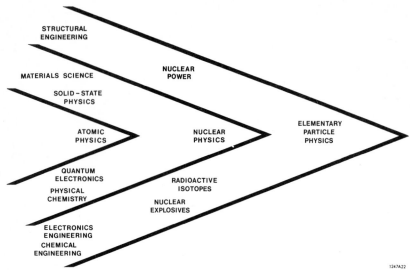

Figure 1-9. Spearhead chart of physics.

From history we have seen how the search for the building blocks of nature has led us from atoms (which turned out not to be so "indivisible") to nuclei and electrons; then from nuclei to neutrons and protons, and now finally (?) to the creation of new unstable particles formed in high-energy collisions. All this has required accelerators of higher and higher energies as we have penetrated deeper into the unknown region of new and unsuspected phenomena. At the same time, the "older" physics has been exploited: Our knowledge of nuclear physics has led to atomic energy, to nuclear explosives, and to isotopes for medicine and industry. Atomic physics is the basis of our entire understanding of chemistry, of the science of materials, and of electronic devices. *Figure 1-9* shows graphically how high-energy particle physics thus constitutes the "spearhead" of our penetration into the unknown.

CHAPTER 2
(R. H. Dalitz)

The Families of Particles

In this lecture, we wish to introduce you to the properties of the "elementary particles", as they are generally called, a name which reflects the notion that these objects represent matter in its most fundamental form. In fact, these particles are now very many in number, certainly more than a hundred of them are known, and there are now very few of us today who feel that there is anything elementary or fundamental about any of them. We know some of these particles much better than others, but this is generally the result of accidental factors which determine the ease with which a particular particle may be studied. We shall find that these "elementary particles" can be classified into a hierarchy of families within families, which reflects their characteristic properties. We shall endeavour to indicate the high degree of order which appears to hold over a wide range of elementary particle systems and of high-energy phenomena, and which suggests that if there are any "basic building blocks" from which matter is formed, they are likely to be rather different from any of the particles we have yet observed. These considerations will lead us on to the final lectures where we shall consider the outstanding open questions about "elementary particles" today, and which will be complemented by Professor Panofsky's final lecture about the experimental tools which are now being developed, whose use may lead us to the answers to these questions in the years ahead.

Before we look at the table of "elementary particles", let us first remind ourselves about several points from the theory of relativity, which we shall need for use along the way. From the previous lecture you will all recall Einstein's expression for the total energy E_0 which corresponds to the rest mass M for a particle at rest, namely.

$$E_0 = Mc^2, \qquad (2.1)$$

where c denotes the velocity of light in free space. Next, you will all remember that the theory of relativity stresses the fact that the description of a given physical event depends on the state of motion of the observer, *i.e.*, that the description of the event cannot be given absolutely but only *relative* to a specified observer. However, the physically interesting facts about such an event are generally those which are *invariant* from observer to observer, *i.e.*, the same for all observers; for example, the cause-effect relationship between two connected events must always remain the same. A particular example of a quantitative invariant is the quantity $(E^2 — c^2P^2)$, where E denotes the total energy of a given system, and P denotes its total momentum. If we consider this invariant for the particular observer whose reference frame is moving with the system in such a way that the total momentum of the system is zero according to his observations, the total energy being E_0, then it clearly has the value $E_0{}^2$, and we have

$$E^2 = E_0{}^2 + c^2P^2, \qquad (2.2)$$

This particular reference frame is usually called the *barycentric frame*. If the system consists of one particle only, then we have $E_0 = Mc^2$, and so the energy-momentum relationship for this particle is

$$E^2 = M^2c^4 + c^2P^2, \qquad (2.3)$$

in any reference frame. The total energy of any particle with mass M and momentum P is given by $\sqrt{(M^2c^4 + c^2P^2)}$. In any reaction process the total sum of this energy for all the particles in a system must have the same value after the reaction as it had before.

We should note that Eq. (2.3) has two solutions,

$$E = \pm \sqrt{(M^2c^4 + c^2P^2)}.$$

We can illustrate the meaning of the negative solution by considering the energy E of the particle (electric charge e) in a potential field V, in which case the Eq. (2.3) takes the form

$$(E — eV)^2 = M^2c^4 + c^2P^2. \qquad (2.4)$$

If we write $E = -\epsilon$, where ϵ is positive, then the left-hand side of this equation becomes $(-\epsilon - eV)^2 = (\epsilon + eV)^2$. In other words, the equation (2.4) may be considered as describing a particle of positive energy ϵ, but with negative electric charge $-e$, moving in the field V. It is characteristic of special relativity that one always has these two kinds of solution; for every physical solution for a particle of mass M and charge e, there is a *charge-conjugate* solution describing a particle of mass M and charge $-e$. In fact, this relationship does not depend on the particle having charge; there are corresponding pairs of solutions also for neutral particles (although there is then also the possibility that the conjugate particle may be identical with the particle itself — π^0 and η provide examples of this situation, in Table I). The two states are referred to as particle and antiparticle, which one is the particle being a matter of convention. The conceptual operation of replacing every particle in a given system or process by its antiparticle is frequently of interest, and is referred to as *particle-antiparticle conjugation,* or more loosely, as *charge reflection*; we shall denote this operation by the symbol C.

Now let us look at Table I. The particles listed here are those which have rather long lifetimes, so we generally refer to them as semi-stable; for this reason, these are really the particles we know best. When is a lifetime regarded as long? — We note that most of these particles have an exceedingly fleeting existence indeed, as measured on our everyday clock. Well, you will see that our energy scale in this Table is MeV (millions of electron volts), with values running up to more than 1000 MeV. One of Heisenberg's uncertainty relations concerns the time interval $\triangle t$ over which a deviation from energy conservation by ϵ can occur, the relation being

$$\triangle t \approx \hbar/\epsilon. \qquad (2.5)$$

A value $\epsilon \sim 1$ *MeV* corresponds to a time interval of $6 \cdot 6 \times 10^{-22}$ sec, and we may regard this as generally a long characteristic time for processes in the nuclear domain; apart from a few exceptional cases where neutral particles may decay by photon emission (and even these have lifetimes of order 100 times this characteristic time), the particle decay lifetimes listed are typically more than 10^{12} times longer than this characteristic nuclear time.

These "elementary particles" separate into three broad families:

(i) *Hadrons:* This name stems from the Greek word for "strong". This class simply consists of all those particles which experience

BARYONS

Name	Y	Mass (MeV)	Free lifetime (sec.)	Dominant Decay Modes
Spin-parity ($\frac{1}{2}+$)				
P	+1	938.3	Stable	—
N	+1	939.6	930	$pe^-\bar{\nu}_e$
Λ	0	1115.6	2.5×10^{-10}	$p\pi^-$ (65%) / $n\pi^0$ (35%)
Σ^+	0	1189.4	0.8×10^{-10}	$p\pi^0$ (53%) / $n\pi^+$ (47%)
Σ^0	0	1192.5	about 10^{-19}	$\Lambda\gamma$
Σ^-	0	1197.3	1.6×10^{-10}	$n\pi^-$
Ξ^0	−1	1315(±0.7)	3×10^{-10}	$\Lambda\pi^0$
Ξ^-	−1	1321.3(±0.2)	1.7×10^{-10}	$\Lambda\pi^-$
Spin-parity ($\frac{3}{2}+$)				
Ω^-	−2	1672.4 (±0.6)	1.3 (±0.4)$\times 10^{-10}$	ΛK^- and $\Xi\pi$

LEPTONS

Name	Y	Mass (MeV)	Free lifetime (sec.)	Dominant Decay Modes
Electronic ($N_e = +1$)				
e^-	—	0.51	Stable	—
ν_e	—	nil	Stable	—
Muonic ($N_\mu = +1$)				
μ^-	—	105.7	2.2×10^{-6}	$e^-\bar{\nu}_e\nu_\mu$
ν_μ	—	nil	Stable	—

MESONS

Name	Y	Mass (MeV)	Free lifetime (sec.)	Dominant Decay Modes
Spin-parity ($0-$)				
$\pi^\pm$	0	139.6	2.6×10^{-8}	$\mu^+\nu_\mu$ or $\mu^-\bar{\nu}_\mu$
π^0	0	135.0	9×10^{-17}	$\gamma\gamma$
$K^\pm$	±1	493.8	1.2×10^{-8}	$\mu^+\nu_\mu$ or (64%) / $\pi^\pm\pi^0$ (21%) / $\pi^\pm\pi\pi$ (7.3%) / $\pi^0\mu\nu_\mu$ (3.2%) / $\pi^0 e\nu_e$ (4.9%)
K^0_S	(±1)	497.8	0.9×10^{-10}	$\pi^+\pi^-$ (68%) / $\pi^0\pi^0$ (32%)
K^0_L	(±1)	497.8	5×10^{-8}	$\pi\mu\nu_\mu$ (28%) / $\pi e\nu_e$ (38%) / $\pi\pi\pi$ (34%)
η	0	549(±0.6)	2.5×10^{-19}	$\gamma\gamma$ (38%) / $\pi\pi\pi$ (53%)

PHOTON

Name	Y	Mass (MeV)	Free lifetime (sec.)	Dominant Decay Modes
γ	0	nil	Stable	—

Table I. The properties of the known stable and semi-stable "elementary particles". Three classes of particle occur—the hadrons (baryons and mesons), the leptons, and the photon.

strong, nuclear interactions, here the baryons and the mesons. These are by far the majority of all the particles known.

The hadrons we know best are the proton (P) and the neutron (N), which are the particles constituting all atomic nuclei. These two hadrons are referred to as the *nucleons*, and they form a subclass of the *baryons*, which are themselves a subclass of the hadrons — the term baryon is a descriptive one, derived from the Greek word for "heavy". A nucleus with charge Ze and mass number A (A = number of nucleons contained) thus contains Z protons and ($A-Z$) neutrons. In particular, the proton is the nucleus of the hydrogen atom; the nucleus of "heavy hydrogen" consists of a neutron and a proton bound together, the di-baryon known as the deuteron (D).

One really outstanding feature of the proton is its stability. Many lighter particles are listed in Table I, and one naturally enquires why the proton does not get rid of its large rest-mass energy by disintegrating into some number of (for example) electrons and positrons. The fact that it does not is very well-known to us, since we would not be here if nuclei could disintegrate in such a fashion. The proton is so stable that no more than 1 in 10^{28} protons disintegrate in the course of a year, and there is no evidence that such disintegration ever occurs. The free neutron does decay slowly through beta-emission, but this process leads to a final proton, so that the nucleon has been transformed rather than disintegrated. We may express this observation as a physical law, "the total number of nucleons remains constant through any interaction process". However, the decay processes for all the baryons listed on Table I lead finally to the emission of one nucleon, and this law has to be generalised to include all the baryons. The simplest way to express this is by the introduction of a quantum number B, called baryon *number*, and by the statement of a *selection rule*.

$$\Delta B = 0 \qquad (2.6)$$

for all physical processes. This selection rule does not explain the stability of the proton; it simply expresses this physical fact in a form which makes it easier for us to trace its implications.

We are already well familiar with another selection rule of this kind, the law of conservation of charge Q (measured in units of the proton charge), which we are perhaps inclined to take for granted because of our familiarity with it. It may be expressed as

$$\Delta Q = 0 \qquad (2.7)$$

for all physical processes.

The antibaryons have values exactly opposite those for the corresponding baryons; thus, for the antiproton we have $Q = -1$, $B = -1$, whereas $Q = +1$, $B = +1$ holds for the proton. The selection rule (2.6) then takes into account the experimental fact that the formation of an antibaryon requires at the same time the formation of some baryon; for example, an allowed production reaction for the antiproton is

$$P + P \rightarrow P + P + P + \bar{P} \qquad (2.8)$$

(ii) *Leptons.* This means essentially "lightweight particles". The lepton we have known longest is the electron e^-, the "carrier of electricity". It has charge $Q = -1$, and $B = 0$ (by definition, these particles have no strong, nuclear interactions). It has a neutral mate, the neutrino v_e ("little neutral one"); its antiparticle, the antineutrino $\bar{v}_e$, is emitted with an electron in the nuclear beta-decay processes,

$$\text{Nucleus } (Z, A) \rightarrow \text{Nucleus } (Z + 1, A) + e^- + \bar{v}_e. \qquad (2.9)$$

Its existence was first postulated by Pauli to account for the fact that the energies of the electrons emitted in a given beta-decay process did not give a line spectrum, but followed a continuous energy spectrum, the interpretation being that the missing energy in the beta emission was due to the simultaneous emission of an energetic antineutrino. We believe that this neutrino has zero mass, but the empirical limit is only that its mass is less than 1/10,000th of the electron mass. By convention, the electron is regarded as the particle, its neutral counterpart being the neutrino v_e; the electronic antileptons are e^+ and $\bar{v}_e$, and both have electron number $N_e = -1$.

There are also muonic leptons, the μ^- meson (with mass $105 \cdot 7$ *MeV*, not so lightweight!) and its neutral counterpart, v_μ. The mass of v_μ is believed to be zero, also, although the empirical evidence only puts an upper limit of 2 *MeV* on this mass at present. The corresponding antileptons ($N_\mu = -1$) are then μ^+ and $\bar{v}_\mu$. The muons are unstable particles, which decay by beta-decay to their neutral counterpart; for example,

$$\mu^- \rightarrow v_\mu + (e^- + \bar{v}_e) \qquad (2.10)$$

They are the "penetrating particles" in cosmic rays. Owing to their large mass, their scattering and emission of bremsstrahlung are very much less probable than these processes are for high energy electrons; since they have no strong nuclear interactions, they are not scattered or absorbed by interactions with the nuclei of the medium along their path. Also, they have a relatively long lifetime, so that their attenuation due to decay generally corresponds to a mean distance measured in kilometres.

It is very remarkable that there should exist two distinct neutrinos, but the evidence about this is quite unambiguous. We shall return to this in a later chapter.

(iii) *The photon.* The quantum of the electromagnetic field stands in a class by itself. We know that it has zero mass, to an exceedingly high degree of accuracy. If the photon had mass m, then its emission by a charged particle would violate energy conservation by an amount of order mc^2. Heisenberg's uncertainty relation allows this to occur only for a time of order $\hbar/mc^2$, and in this time the photon can travel a distance $c(\hbar/mc^2) = \hbar/mc$. This means that the electromagnetic field would be of finite range, proportional to the form

$$Q_1 Q_2 \exp\left(- mcr/\hbar\right)/r, \tag{2.11}$$

for the interaction between two charges Q_1, Q_2, both at rest. We know from macroscopic electrical experiments that the potential between two charges has the Coulomb form $Q_1 Q_2/r$, to a very high degree of accuracy. These experiments serve to show that the photon mass could not exceed 10^{-21} of the electron mass, so that we are rather inclined to accept that $m = 0$ for the photon.

It is these Coulomb forces of attraction between electrons and nuclei which give rise to atoms, and then to molecules, which are of course big clusters of atoms, involving many nuclei in general, the electron clouds encompassing the whole system, binding the atoms together. The electrons have discrete states of motion in these atoms and molecules, as first became established from the observation that the light emitted was in discrete quanta with a definite frequency ν_{ab} characteristic of the two states, the energy of the photon being $\hbar\nu_{ab}$ (ν_{ab} measured in radians per sec here) given by

$$\hbar\nu_{ab} = E_a - E_b \tag{2.12}$$

where E_a and E_b measure the energies of the two states. It was from observations of this kind that the quantum theory came into being in the 1920's.

We can make a rough estimate of the energies typical of chemical forces as follows. The last electron of a given neutral atom sees a Coulomb field appropriate to charge $Z = +1$. If the size of the atom is R, then Heisenberg's uncertainty relation indicates that the electron momentum in this atom is of order $\hbar/R$. Hence the electron has kinetic energy of order $(\hbar/R)^2/2m$ and potential energy roughly $-e^2/R$. The most favourable situation corresponds to the minimum in the total energy

$$E(R) = -e^2/R + (\hbar/R)^2/2m, \qquad (2.13)$$

which occurs at $R = \hbar^2/(me^2) \approx 10^{-8}$ cm. The energy value is $-(e^2/\hbar c)^2 mc^2/2 \approx -10\ eV$ since $e^2/\hbar c$ is the fine structure constant $\alpha = 1/137$ and $mc^2 = 0.51\ MeV$ for an electron. These quantities account for the sizes typical for atoms, and for the

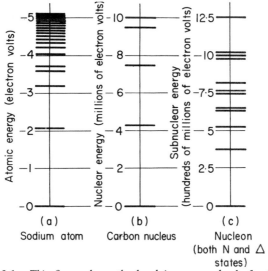

Figure 2-1. This figure shows the low-lying energy levels for (a) the sodium atom, (b) the carbon nucleus C^{12}, (c) the nucleon, including both the $I = \frac{1}{2}$ and $I = 3/2$ excited states. Notice the difference in energy scales, which are electron volts for the atomic levels (a), MeV for the nuclear levels (b), and 100 MeV for the sub-nuclear excitations. The atomic levels are distinctive in that they crowd towards a continuum limit at 5.1eV; this is characteristic of a system whose internal forces are of long range, as is the case for the coulomb interaction between electric charges.

typical energy involved in their interactions; we may compare this energy with the energy for the first excited state in the *Na* atom, given on *Figure 2-1*. Given the nuclei, and the electrons, then it is to be expected that essentially all of the facts of chemistry should be a consequence of the Coulomb interactions between these particles, although it must be admitted that, except for the simplest systems, the calculations are immensely involved and largely beyond our calculational powers today.

We turn next to the atomic nuclei. From the strong scattering he observed at large angles for alpha particles incident on atoms, Rutherford deduced that the Coulomb force on the alpha particle followed the form $2Ze^2/r^2$ to very small distances, where the forces were so strong that they gave rise even to large backward scattering. With alpha particles of higher energy, it is finally possible to reach the nuclear surface, as shown by the deviations from Coulomb scattering then found and by the nuclear reaction processes induced. This leads to the estimate of about $A^{1/3} \times 1 \cdot 3 \times 10^{-13}$ *cm* for this radius, so that the nuclear volume is roughly proportional to the number of nucleons, a property referred to as the saturation of nuclear forces. This size corresponds to a mean distance d between two nucleons in the nucleus, of about $1 \cdot 6 \times 10^{-13}$ *cm*, which indicates that the range of nuclear forces is to be expressed with unit 10^{-13} *cm* = 1 *fm*. We can make a rough dimensional argument here, parallel to the discussion following Eq. (2.13), replacing m by the nucleon mass M and e^2 by a corresponding factor g^2 for the interaction between nucleons. The mean distance between nucleons is to be identified with $\hbar^2/Mg^2$; this gives the estimate $g^2/\hbar c \approx 1/8$, a force strength about 20 times that for the Coulomb interaction, and a characteristic energy

$$(g^2/\hbar c)^2(Mc^2/2) \approx (1/8)^2(470) \approx 7 \ MeV, \qquad (2.14)$$

which may be compared with the energy scale given in *Figure 2-1* for the level system for the C^{12} nucleus.

The detailed nuclear forces between two nucleons have been deduced in some detail from the analyses of *PP* and *NP* elastic scattering, as function of energy. The *PP* and *NN* forces have been compared rather directly by comparing the properties of mirror nuclei, such as $He^3 = PPN$ and $H^3 = NNP$, the first being related to the second by the operation of interchanging N and P.

This evidence all shows that, for a given state of motion, the nuclear force between two nucleons is essentially the same no matter which pair of nucleons, *PP, PN,* or *NN,* are considered. In other words, to a good approximation, the nuclear forces are *charge-independent,* apart from the Coulomb forces due to their charge or internal charge distribution. This explains the closeness of the *P* and *N* mass values. They represent a *charge doublet,* two states of the nucleon, the mass difference of 1·3 *MeV* between them being attributed to the fact that one particle is charged, the other not.

The energy levels for the C^{12} nucleus are shown on *Figure 2-1 (b).* It is quite comparable with that for the *Na* atom except that (i) the nuclear energy scale is about 10^5 greater than that for the atomic energy, and (ii) the atomic levels show a crowding as the excitation energy approaches the ionization limit; this latter feature is characteristic of an interaction with long range, such as

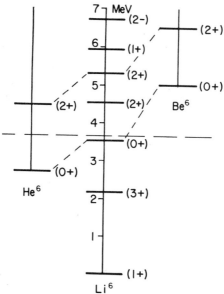

Figure 2-2. The energy levels are shown for the nuclear systems with mass number A = 6. The dotted lines connect the corresponding levels in He^6, Li^6, and Be^6, which are therefore charge triplet states. All of the other states given are charge singlet. The horizontal dashed line marks the level of energy corresponding to (He^4 + 2 nucleons) for each charge value.

the Coulomb interaction, and is therefore naturally absent in the nuclear case. Another set of nuclear energy levels is shown in *Figure 2-2* for the $A = 6$ nuclei, He^6, Li^6 and Be^6. We note that there are levels in Li^6 which correspond to the levels of He^6 and Be^6. For example, the $J = (0 +)$ level at $3 \cdot 56$ *MeV* excitation in Li^6 corresponds to the ground states of He^6 and Be^6 (the progressive decrease in binding energy for the $A = 6$ nuclei as Z increases from 2 to 4 is due to the increasingly effective Coulomb repulsions). Again the first excited states for He^6 and Be^6 have spin $(2 +)$ and excitation energies $1 \cdot 8$ and $1 \cdot 5$ *MeV*, respectively; the corresponding level in Li^6 lies at $5 \cdot 36$ *MeV*, i.e., $1 \cdot 8$ *MeV* above the $(0 +)$ level at $3 \cdot 56$ *MeV*. We therefore describe these levels as forming charge triplets; all the other levels in Li^6 below $5 \cdot 36$ *MeV* are without counterparts in He^6 and Be^6, so that they are charge singlet states. All the states known for the $A = 6$ nuclei appear as charge singlets in Li^6 or as charge triplets centred on Li^6.

This feature of the nuclear energy levels for $A = 6$ holds quite generally for the set of nuclei with the same mass number A, at least until the mass number is so large that the effects of the Coulomb repulsions between the Z protons begins to play a major role in the nuclear energy. For a given value of A, the nuclear levels appear as *charge multiplets*, a set of states with the same spin and parity forming a multiplet of $(2I + 1)$ levels, with essentially the same binding energy (relative to all the constituent neutrons and protons) after allowance for the energy of Coulomb repulsion between the protons. The number I is called the isospin of the charge multiplet, and the $(2I + 1)$ equivalent states are labelled by an "isospin component" I_3 which runs through the values $- I, - I + 1, \ldots + I$ with increasing charge for the nucleus. The charge multiplet is centred on the nucleus with $Z = A/2$ if A is even or on the pair of nuclei with $Z = (A \pm 1)/2$ if A is odd. The nucleons form a charge doublet, with $I_3 = - 1/2$ for N and $I_3 = + 1/2$ for P; so also do the mirror nuclei H^3 and He^3. Generally, the charge of the nucleus specified by these quantum numbers (I, I_3) is given by

$$Q = I_3 + A/2, \qquad (2.15)$$

since this clearly holds true for the nucleons themselves (for which case $A = 1$). The charge multiplet states are frequently referred to as *isobaric states* since they would have exactly the same mass if all the electromagnetic effects (N–P mass difference and P–P Coulomb repulsions) could be abolished. In fact, as I_3 increases, the number of protons in the nucleus increases and their Coulomb repulsion ensures that the nuclear state with $I_3 = +1$ has the highest energy, in fact an energy so high that the state may be unstable with respect to the emission of one or two protons. This last situation holds even for Be^6, depicted in *Figure 2-2*, since its ground state lies above the threshold energy for $PPHe^4$; however, the Be^6 ground state and first excited state correspond to rather well-defined resonances in the system ($PPHe^4$) and so it is reasonable to compare them with the corresponding states in He^6 and Li^6 (some of which are similarly unstable but well defined in energy). Even for rather large nuclei, with mass number A of order 100, where the ground states observed have large negative values for I_3 (i.e., many more neutrons than protons), the corresponding isobaric state in the neighbouring nucleus can still be picked out quite clearly from observations on reactions such as the (P, N) exchange reaction,

$$P + \text{Nucleus } (Z, A) \rightarrow N + \text{Nucleus } (Z + 1, A)^* \quad (2.16)$$

This phenomenon of isobaric charge multiplets in nuclei stems from the property of charge independence for the nuclear forces, and provides most striking evidence for it. This property corresponds to the existence of a special symmetry in the nuclear forces, its technical name being SU(2) symmetry, which states essentially that, as far as nuclear forces are concerned, the basic constituents of nuclei, N and P, are completely equivalent objects. Only when the much weaker forces of electromagnetism, characterised by the parameter $e^2/\hbar c \approx 1/137$ rather than the parameter $g^2/\hbar c \approx 1/10$ appropriate to nuclear forces, come into play, does any distinction arise between N and P. In fact, we may say that the outstanding role of the electromagnetic field is to provide a means for distinguishing between the two states of the nucleon.

At one point in time, then, it appeared rather hopeful that physicists could account for all the properties of matter in terms of the particles P, N, e^- and γ. The protons and neutrons bound together to form all the various nuclei. The photon (electromagnetic) field generated the Coulomb attraction between the electrons and the nuclei, leading to atomic states, and also accounted for all the electromagnetic radiations (as diverse as radio waves, light and nuclear gamma rays); the forces between atoms also resulted from the Coulomb repulsions and attractions and could account for the whole of chemistry, at least in principle. These photons were emitted discretely in the transitions between the quantum energy levels of the molecules, of the atoms or of the nuclei. The beta-decay process (2.9) occurred for many otherwise stable nuclei and required the existence of the neutrino, a particle which appeared to play no other role in physics, and many other nuclei decayed by α-emission, a process which could be understood quantitatively as a straightforward disintegration process, generally slowed down by a very large factor since the α-particle ($= He^4 = PPNN$) had to leak quantum-mechanically through the strong potential barrier due to the Coulomb repulsion between the α-particle and the residual nucleus, the α-particle having insufficient energy to surmount this potential barrier. All that seemed left to do was to determine more precisely the detailed properties of the nuclear forces and to use them to explain the energy level structure of nuclei and the $\alpha-$, and $\beta-$ and $\gamma-$ transitions observed between them.

In 1935, Yukawa suggested that the nuclear forces could be due to the existence of an intermediate meson field which interacted with the nucleons, rather in the same way that the photon field generates the Coulomb force between charged particles. There were two important differences. First, the nuclear forces were known to be of short range, comparable with $1 \cdot 6$ fm as we argued above. A comparison with the potential form (2.11) indicated that the meson field quanta would have to be massive, with a mass of the general order of 100 to 200 MeV. Second, there must be meson field quanta carrying charge; this was required to account for the saturation properties of the nuclear forces, mentioned above. Since the muon was known and had mass 105 MeV, it was first thought possible that this might be the required meson. However, it was found that the μ^- meson could be captured into atomic orbits about atomic nuclei and could remain there, its motion overlapping the nuclear region rather strongly, for a time comparable with its lifetime for free decay. Although the muons appeared copiously in the cosmic radiation, they had no interactions with nuclei stronger than those due to the Coulomb forces. Marshak and Bethe finally suggested that these muons must be the decay products resulting from the decay of some heavier meson which did have strong nuclear interactions, and it proved to be so. In 1947, at Bristol, Powell and Occhialini found an event which showed clearly the emission of a muon in the decay of a cosmic ray particle which had come to rest in their nuclear emulsions. This particle is now known as the π-meson (or "pion") and the decay process was established to be

$$\pi^+ \rightarrow \mu^+ + \nu_\mu, \qquad (2.17)$$

with lifetime comparable with 10^{-8} sec. These pions proved to be the "nuclear force mesons" required by Yukawa's proposal, their mass being about 140 MeV and their coupling strength being comparable with the required value $g^2/\hbar c \sim 0\cdot1$. As shown on Table I, the pions form a charge triplet, and have charge-independent nuclear interactions. This fact provides the simplest interpretation for the charge-independence of nuclear forces, that they are due to an intermediary pion field whose interactions are charge-independent.

The antiparticles to the pions are the pions themselves. Under the operation C (particle-antiparticle conjugation), we have $\pi^\pm \rightleftharpoons \pi^\mp$, the π^0 meson being its own antiparticle. The π^0 meson has a rather rapid decay (lifetime of order 10^{-16} sec) through electromagnetic interactions, $\pi^0 \rightarrow \gamma\gamma$; it also has the visible decay mode (frequency about 1%), $\pi^0 \rightarrow \gamma \, e^+e^-$. The mass difference of about 4.6 MeV between $\pi^\pm$ and π^0 is attributed to the effect of the electromagnetic interactions effective in the $\pi^\pm$ system. Indeed, we shall take the mass of 5 MeV as a typical measure of the strength of the electromagnetic interactions in the subnuclear (or "elementary particle") domain.

At last, all the particles and fields necessary to account for all nuclear, atomic and molecular phenomena appeared established; only the details were left to be worked out, all the essential ingredients were at hand. Alas, already within one year from the discovery of the pion, Powell's group at Bristol reported the observation of a much heavier meson, the K^+ meson with mass about 500 MeV, identified from measurements on its three-particle decay mode $K^+ \rightarrow \pi^+ \pi^+ \pi^-$, from rest in nuclear emulsion, and Blackett's group at Manchester reported the first evidence for V-particles, the

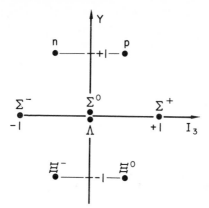

Figure 2-3. The octet of baryon states are displayed on a (I_3, Y) plane, each particle being marked by a corresponding point.

name given to the characteristic V-shaped pair of tracks seen in a cloud chamber when a neutral particle decays into two oppositely-charged particles. The V-particle phenomena turned out to be quite complex, owing to the large number of decay modes listed in Table I which are of this type. The first to be identified was a baryon, the Λ-particle with decay mode $\Lambda \to P\pi^-$; another strong contributor was the neutral K-meson, with decay mode $K_S^0 \to \pi^+\pi^-$. There were also found charged V-particles, which were seen as a sharp kink in the track of a charged particle in the cloud chamber. These included the processes $\Sigma^+ \to P\pi^0$ and $K^+ \to \pi^+\pi^0$ and $\mu^+ \nu_\mu$. Although these events were not common, they were produced quite strongly, at the rate of about 1% of the pions produced in the cosmic ray interactions, so that their production definitely occurred through interaction processes of nuclear strength. These heavier mesons and baryons could not be ignored, and it became clear at once that the physical situation concerning the "elementary constituents of matter" on the nuclear level was much more complicated than physicists had suspected. However, the precise disentangling of these phenomena had to wait until the large proton accelerators were constructed in America, at Brookhaven and Berkeley, and made available a copious supply of these new particles for study under controlled laboratory conditions.

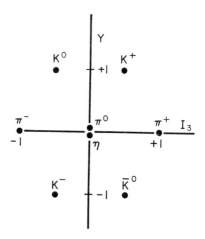

Figure 2-4. The octet of mesons with spin-parity (0—) are displayed on a (I_3, Y) plane, each meson being represented by a dot at the appropriate location.

Now let us look at Table I and *Figures 2-3* and *2-4*. We see that all of the baryons appear grouped in charge multiplets, the nucleon doublet, the Λ singlet, the Σ triplet and the Ξ doublet. The same holds for the meson states. Clearly the property of charge independence holds also for the strong nuclear interactions for all these hadrons, as was first suggested by Gell-Mann. But it is the long lifetimes listed for these hadrons in Table I which are the most puzzling items. Apart from the few cases of electromagnetic decay, the lifetimes are at least 10^{13} times longer than a typical nuclear time (which we may estimate here as $\hbar/(100 \ MeV) \approx 10^{-23}$ sec, since the typical energy differences are of order 100 MeV), whereas the frequency of production observed for these particles requires that they should have some interactions of nuclear strength. This discrepancy suggested that there was some selection rule operating. Gell-Mann and Nishijima independently made suggestions equivalent to the introduction of a new additive quantum number, the *strangeness* number s. This name was chosen because it referred to these new, heavier particles, which were referred to as "strange particles" in the early years when they were first being explored. This is a most unsuitable name for them now that we have grown rather accustomed to

their study — we are now more inclined to stress their relationships of similarity with respect to the particles we have known
longer.

The value $s = 0$ is given to the nucleons and to the pions,
and we suppose that there is a conservation law for strangeness,

$$\Delta s = 0, \tag{2.18}$$

for all strong nuclear processes, i.e., the sum of the strangeness
numbers for all particles after a strong nuclear interaction must
be the same as this sum for all the particles before this reaction.
If we assign $s = -1$ to the Λ hyperon, then the observed
decay process

$$\Lambda \rightarrow P + \pi^-$$
Strangeness: -1 $0 + 0$ $\tag{2.19}$

is forbidden to occur through the strong interactions. To avoid
the possibility of a rapid electromagnetic decay, $\Lambda \rightarrow N\gamma$, we must
assume that the conservation law (2.18) holds also for the electromagnetic interactions. Since we do observe the decay reaction
(2.19), we have to conclude that the weak decay interactions
do violate strangeness conservation (although only by effects of
order 10^{-13}). For this reaction, the strangeness change is actually
$\Delta s = +1$.

About 1953, the production process for the Λ particle was first
established for π-proton collisions,

$$\pi^- + P \rightarrow \Lambda + K^0, \tag{2.20}$$

a process of associated production. Strangeness conservation then
requires $s = +1$ for the K^0 meson, and therefore also for the
other member of the K doublet, the K^+ meson. This assignment
requires the strangeness change $\Delta s = -1$ for any of the decay
processes listed for the (K^0, K^+) doublet in Table I, so that those
can be due only to the weak decay interactions, as the lifetime
observed indicates must be the case. The Σ production reactions,
for example

$$\pi^+ + P \rightarrow \Sigma^+ + K^+ \tag{2.21}$$

require the assignment $s = -1$ for the Σ particles; this is also
indicated by the short lifetime for the decay process $\Sigma^0 \rightarrow \Lambda\gamma$.
The assignment $s = -2$ is required for the Ξ particles, in order
to account for the decays $\Xi \rightarrow \Lambda\pi$, observed to be slow, with the
rule $\Delta s = +1$. This assignment is confirmed by the production

process observed to occur for the Ξ hyperon,

$$\pi^- + P \to \Xi^- + K^+ + K^0, \qquad (2.22)$$

involving the simultaneous production of two K-mesons with the Ξ hyperon.

This $\Delta s = 0$ rule for strong and electromagnetic interactions has been tested very widely in high-energy reaction processes. It is satisfied by all the reactions which are observed to occur through strong or electromagnetic processes, an example of the latter being

$$\gamma + P \to \Sigma^+ + K^0. \qquad (2.23)$$

On the other hand, reactions which violate $\Delta s = 0$ are not observed to occur strongly. Two examples of this remark are as follows: First, the reaction

$$N + N \to \Lambda + \Lambda \qquad (2.24)$$

has a particularly low threshold (it required neutrons of more than 770 *MeV* laboratory kinetic energy incident on a target containing neutrons). Despite favourable conditions, this reaction has never been detected. Second, a Λ particle in contact with nucleons is semi-stable since, for mass number A, the system of one Λ particle and $(A - 1)$ nucleons in its ground state is the system with lowest energy for baryon number $B = A$, charge Q (as appropriate), and strangeness $s = -1$. There is simply no strong reaction process possible for this system, consistent with the B, Q and s conservation laws, and the Λ particle will remain stable in this nuclear environment for a time of order 10^{-10} sec, until the Λ particle decays, either in its usual mesonic mode (2.19) or by a non-mesonic mode

$$\Lambda + P \to N + P + 176 \ MeV \qquad (2.25)$$

with a large energy release, which becomes possible through the weak decay interactions owing to the presence of nucleons in the neighbourhood of the Λ particle. Such systems, the Λ-hypernuclei consisting of a Λ particle attached to an ordinary nucleus, are well known and have been studied in some considerable detail because of their direct relevance to the Λ-nucleon nuclear forces; even several examples of $\Lambda\Lambda$-hypernuclei are known, from which it has been possible to obtain rather direct information on the nuclear forces in the $\Lambda\Lambda$ system. On the other hand, the Σ particle generally undergoes a rapid transformation to a Λ particle through strong interactions in the presence of nucleons, for example through the reaction

$$\Sigma^- + P \rightarrow \Lambda + N. \tag{2.26}$$

The same is true for the Ξ particle, its transformation being due to the allowed reaction

$$\Xi^- + P \rightarrow \Lambda + \Lambda \tag{2.27}$$

The K-meson situation looks a little confused on Table I. The K^+ and K^- particles definitely do not belong to the same multiplet. The production processes for the K^+ meson, and the absorption processes for the K^- meson, such as

$$K^- + P \rightarrow \Lambda + \pi, \tag{2.28}$$

show clearly that K^+ and K^- have opposite strangeness, consistent with the relationship expected between particle and antiparticle. Their neutral counterparts, K^0 and $\overline{K}^0$, then have opposite strangeness values and the same mass, insofar as they are particle and antiparticle. However, the weak interactions do not respect strangeness and therefore mix up these two states, with the result that the two neutral K mesons observed, K_s that with the short lifetime and K_L that with the long lifetime, do not have a unique strangeness but each involve a mixture of strangeness $+ 1$ and strangeness $- 1$. The mass difference between K_S and K_L is therefore due only to the weak interactions; it has been measured, giving the tiny difference $(m_L - m_S) c^2 = 3 \cdot 6 \times 10^{-6} \, eV$. The neutral K-meson phenomena are quite complicated and fascinating in themselves, and we shall mention them briefly again in the lecture on weak interactions.

The last semi-stable meson to be mentioned is the η meson.

This is strongly produced in the reactions

$$\pi^- + P \rightarrow N + \eta, \qquad\qquad (2.29a)$$

$$\gamma + P \rightarrow N + \eta, \qquad\qquad (2.29b)$$

for incident energies near the threshold energy for η production. Despite its relatively high mass value (about 550 *MeV*), it has a relatively long lifetime, for the electromagnetic decay process $\eta \rightarrow \gamma\gamma$ represents quite a large fraction of its decay processes.

CHAPTER 3
(W. K. H. Panofsky)

The Interaction and Identification of Particles

In the first lecture we traced the history of man's study of the structure of matter, and we saw why the pursuit of these goals requires higher and higher particle energies. In the second lecture we identified the families of particles as far as we know them today. We now turn to a third topic: How can the experimental physicist tell the different particles apart? What properties do they exhibit in their interaction with matter or other behaviour that allow us to identify them?

In brief, the particle properties we have to work with are the following:

1. Charge ($+$, — or neutral).

2. Momentum or $\vec{p}$ (as measured by the curvature of a particle trajectory in a magnetic field).

3. Velocity $\vec{v}$ (as measured by time-of-flight, by ionization or similar effects, and by observation of "Cerenkov light").

4. Range and multiple scattering in matter.

5. The probability of nuclear interaction in matter.

6. The probability of producing various specific products of interaction in matter.

7. The life-time and spontaneous decay mechanism of the particle.

Clearly many lectures could be dedicated to these subjects; our purpose here will be simply to outline these topics to give you a better understanding of the operation of various particle detectors and experimental arrangements.

The dominant interaction of *charged* particles passing through matter is *ionization*, i.e., the ejection of electrons from atoms near the particle's trajectory. The mechanism is as follows: A charged particle moving with velocity v passes in the neighbourhood of an atom (*Figure 3-1*). As the result of the passage one can easily calculate from the "inverse square law" (Coulomb's law) that an atomic electron at a distance b will receive a transverse momentum Δp as given by

$$\Delta p = \frac{2e_1e_2}{bv} \tag{3.1}$$

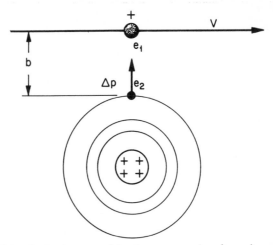

Figure 3-1. Ionization caused by the passage of a charged particle.

where e_1 and e_2 are the charges on the particle and the atomic electron respectively. If this "kick" given to the electron is large enough, and occurs rapidly enough, then the atomic electron will be ejected; as a result, the incident particle suffers a loss in energy. One can average the probability of this process over various values of b and over the energies with which the atomic electrons are bound to the nucleus and can therefore derive an *average* energy loss per unit length (dE/dx) by ionization. This quantity depends only on the charge e_1 of the incident particle and its velocity v; the mass or other characteristics of the incident particle matter very little. As the result of this averaging process one obtains typical curves like the one shown in *Figure 3-2*, plotted for aluminum.

The detailed structure of the stopping material does not matter much either; what is important is the number of electrons/cm^2 in the path of the beam. Since for each electron in the atom there is one proton in the nucleus, and since for all nuclei (except hydrogen) there are roughly as many neutrons as protons, the number of electrons/cm^2 and the number of grams/cm^2 are roughly proportional.* More accurate curves like the ones shown in *Figure 3-2* have been computed for all important materials.

---

* For this reason it is customary in particle physics to measure thickness of absorbing materials in g/cm^2; this is the thickness in *cm* times the density in g/cm^3.

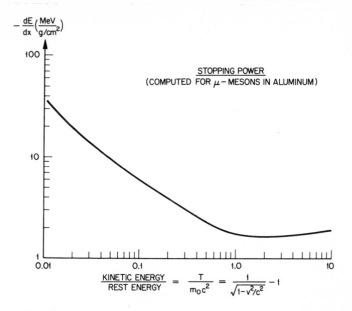

Figure 3-2. Average ionization energy loss curve for aluminum.

We note the following general features:
1. For particles of kinetic energy well below the rest mass the ionization falls rapidly with energy.
2. For "relativistic" particles, i.e., particles of high energy relative to their rest energy, the ionization is almost constant; at the point of "minimum ionization" the energy loss for a singly charged particle is about 2 MeV/(g/cm^2).
3. At low energies ionization is a fair measure of the velocity of a singly charged particle; at high energies it measures only the charge.

Ionization is the factor controlling the *range* of a particle in matter. It also controls the density of a track in the various devices we will discuss later (bubble chamber, cloud chamber, etc.) where particle tracks can be photographed. Finally, in those particle detectors (ion chambers, proportional counters) where the ions produced by the incident particles are collected, the electrical signal measures the energy loss by ionization. In a particle counter in which the traverse of the particle produces a light flash (scintillation counter), the light output is generally proportional to the

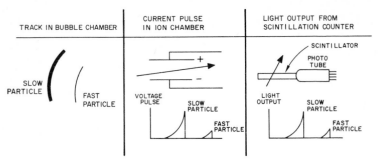

Figure 3-3. Methods of measuring ionization.

energy loss by ionization. These methods by which ionization is measured are shown in *Figure 3-3*. If the energy lost in these detectors is only a small fraction of the total particle energy these methods can distinguish slow and fast particles.

From the energy-loss equation one can compute the range of a particular particle in various materials; these calculations have resulted in tables of range-energy relations. However, at very high energies the range as limited by ionization loss becomes so great that, before most fast particles are stopped by ionization, they have a large chance of producing a nuclear collision.

Since the energy loss by ionization, or "stopping power", as shown in *Figure 3-2* is a function only of the velocity, we can write

$$- \frac{dE}{dX} = f(v); \qquad\qquad 3(.2)$$

the energy is a product of the rest mass m_0 and another function $g(v)$ of the velocity. Therefore the range R is given by

$$R = \int_{E_0}^{0} \frac{dE}{dE/dX} = m_0 \int \frac{g(v)}{f(v)}\, dv = m_0\, F\left(\frac{E_0}{m_0}\right) \qquad (3.3)$$

where F is some other function; this is plotted for aluminum in *Figure 3-4*. Hence for singly charged particles the ratio of range to rest mass is an almost universal function of the ratio of kinetic energy to rest mass. Let us look at some extreme examples: A 4 *MeV* proton (rest mass 938 *MeV/c²*) will penetrate $2 \cdot 8 \times 10^{-2}$ *g/cm³* of aluminum, or about 0·1 millimetres, while the range of a 20 *GeV* μ-meson from an accelerator is 10,000 *g/cm²*, or 36 metres of aluminum (!).

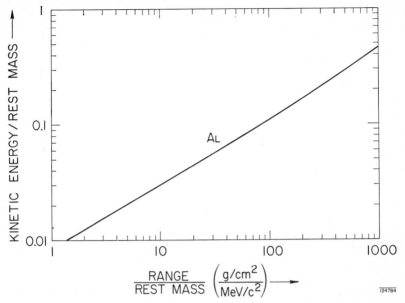

Figure 3-4. Ratio of range to rest mass plot for aluminum.

These examples are correct only if energy loss by *ionizing collisions* is the dominant mechanism of stopping power; this is true in the examples chosen. In the first case (that of the low-energy proton) this is true because the range is short; in the second case it is true because μ-mesons have only negligible interactions with nuclei and therefore in their deep penetration are stopped only by ionizing collisions with electrons. In contrast, almost all very high energy hadrons (protons, pions, neutrons), whether charged or not, are stopped by interactions with the atomic nuclei; as a *rough* estimate such high energy particles interact once for every 50 *g/cm²* (20 *cm* of aluminum) traversed. Thus a 20 *GeV* proton penetrating matter would dissipate its energy long before it was stopped by ionization.

Something else happens when a fast particle penetrates through matter: it scatters. By this we mean that it does not exactly continue in a straight line but as the result of many small collisions the particles "wiggle" their way through matter. It can be shown that after passing through an absorber of thickness X, the initial

direction of the particle *on the average* will be deflected through
an angle δ which is given by

$$\delta = \text{proportional to } (\sqrt{X}/pv) \qquad (3.4)$$

where p is the particle momentum and v its velocity. Hence
measurement of the average scattering gives another handle on
the mass and speed of the particle.

We thus find:

(a) μ-mesons and charged hadrons up to a few hundred *MeV*
    penetrate matter in accordance with the range-energy
    relation as defined by ionization;

(b) hadrons, charged or not, of high energy penetrate until
    they collide with a nucleus in the stopping material.

But what about γ-rays and high-energy electrons? Here the
story is different. You all know that electrons of sufficient energy
hitting a target will emit X-rays (which would be called γ-rays at
MeV energies or above). The reason is that electrons are so
light that their path is deflected by the electric field within each
atom; when a charged particle is accelerated it "radiates" photons
(X-rays or γ-rays). As a reaction we write this as follows:

$$e^- + \text{nucleus} \rightarrow e^- + \text{nucleus} + \gamma \qquad (3.5)$$

The likelihood of this process is larger if the electric field in the
atom is higher, which is the case for the heavier elements where
the nucleus carries more charge. As an example, a fast electron
will lose two-thirds of its energy through radiation after passing
through 6 g/cm^2 (1/2 *cm*) of lead (atomic number 82), while
the same result would occur after penetrating 26 g/cm^2 or about
10 *cm* of aluminum (atomic number 13). Therefore, high-energy
electrons can be distinguished from other charged particles through
their radiation of photons, and their penetration in matter relative
to that of protons is much less for absorbers of high atomic
number but becomes comparable for light materials.

What happens to very high energy photons? The dominant
process is "pair production", i.e., the photon can materialise
(according to Einstein's $E = mc^2$ relation) into a pair of positrons
and electrons according to the reaction

$$\gamma + \text{nucleus} \rightarrow \text{nucleus} + e^+ + e^- \qquad (3.6)$$

In a magnetic field this reaction would look as shown in *Figure 3-5*.
An incident γ-ray strikes a thin plate of heavy material (a "con-

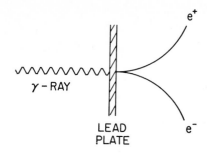

Figure 3-5. Electron-positron pair production.

verter") and a pair of oppositely curved tracks emerges. What is the ultimate fate of this electron pair if the absorber is thick? In that case the electrons and positrons "radiate" according to the discussion above; the resultant γ-rays form pairs again, and so on. The result is a "shower" of particles (*Figure 3-6*). Note that the charged particles multiply rapidly, with the energy of each particle becoming less. How long will this go on? It will continue until the energy of each particle has become so small that it will be less likely to radiate but will then lose energy by the simple ionization mechanism described in the beginning of this chapter. Radiation and ionization energy losses become about equal at 8 *MeV* in lead and at 40 *MeV* in aluminum (as discussed above, radiation is relatively more likely in heavier materials). A shower can, of course, be produced whenever it is started by an electron or by a γ-ray; in either case the energy will eventually be dissipated by the ionization of low-energy electrons.

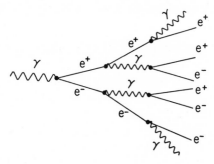

Figure 3-6. Particle shower from electron pair in thick absorber.

At high energies electrons can thus easily be distinguished from muons and protons if a "shower" can be identified; for instance if in a counter the multiplicity of charged particles produced by a single particle can be registered. A high-energy γ-ray can be identified by observing that a neutral (non-ionizing) particle becomes a charged (ionizing) pair of particles after passing through a thin plate made of heavy material.

We have thus far discussed the phenomenon by which penetration into matter results in characteristic "signatures" for energetic hadrons (i.e., neutrons, protons, pions, etc.), muons, electrons and photons. Let us now turn to other phenomena *not* involving penetration through matter. The most important of these is the curvature of the trajectory of a particle of charge e and velocity v in a magnetic field B. The bending force, given by $B \cdot e \cdot v$, is then balanced by the centrifugal force pv/R, where p is the momentum and R the radius of curvature of the resultant orbit (*Figure 3-7*). We thus obtain

$$Bev = pv/R \qquad (3.7)$$

or

$$BR = p/e. \qquad (3.8)$$

Thus measurement of the radius of curvature of the trajectory in a magnetic field measures the *momentum* of the particle. If B is measured in gauss, R in *cm*, and p in eV/c,* this relation becomes numerically

$$BR = p/300 \qquad (3.9)$$

Frequently experimental arrangements permit direct observation of the *velocity* of a fast particle. This can be done by electronic timing in the flight path of the particle to split nanosecond accuracy

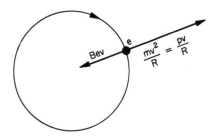

Figure 3-7. Magnetic field bending of charged particle.

(one nanosecond $= 10^{-9}$ sec). Even with this precision velocity measurement becomes less powerful for very high energy particles. Since at the highest energies the speed of all particles, irrespective of their mass, approaches the velocity of light, time-of-flight methods generally become unproductive above energies of several *GeV*.

A very special method of measuring velocity is the observation of "Cerenkov" light, named after its experimental discoverer. Let me explain the principle of this phenomenon: At very high energies the speed of all particles approaches the speed of light *in empty space*. However, the speed of light in a *material medium* is less than the speed in vacuo by a factor *n* called the refractive index; most of you are familiar with this from studies in optics. Therefore an energetic particle passing through, for example, glass can move faster than the speed of light in that medium, i.e., we can have conditions such that $c/n < v$. The situation is similar to that of supersonic flight; an aeroplane can fly faster through air than the speed of sound in that medium: In that case, the result is an acoustic shockwave. Similarly our fast charged particle will produce a conical optical shockwave called "Cerenkov radiation"

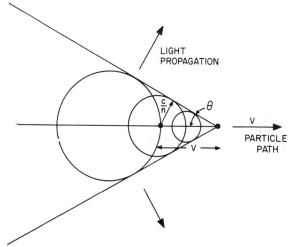

Figure 3-8. Cerenkov radiation showing conical optical shockwave.

---

* The momentum *p* has units of energy divided by velocity; in high-energy physics, it is conventional to measure *p* in units of electron-volts (*eV*) divided by the velocity of light (*c*).

as shown in *Figure 3-8*. If we assume that as the fast particle moves through the medium light is continuously emitted, the light waves will reinforce one another along a conical wavefront; the semi-apex angle ϑ of the cone is given by (see *Figure 3-8*):

$$sin\ \vartheta = \frac{c}{nv} \tag{3.10}$$

which is less than unity if $c/n < v$. Therefore measurement of the angle of light emission gives the value of the particle velocity. Moreover, if the refractive index n is properly chosen in a particular experiment, then one can discriminate among particles of different masses which have passed through a bending magnet: As we have seen previously, the magnet has fixed the *momentum*; thus particles of different mass have different velocities, say v_1 and v_2; if $v_1 > c/n$ while $v_2 < c/n$, only the particle of velocity v_1 will emit a light flash when passing through the medium.

We see from this brief outline that the physicists have many tools which in effect "weigh" the particle, i.e., they determine its mass. These methods have one feature in common: They become harder to apply as the energy becomes higher. There are, of course, other, more specific means to identify a particle, once we know how each particle decays, what its specific reaction products are in collision, or how long it lives. But such information is not available when one searches for new particles. The discovery of a new particle must either rest on a mass measurement or on a measurement of some other physical property; or else its specific behaviour must be shown to be inconsistent with that of other particles known before.

CHAPTER 4
(W. K. H. Panofsky)

The Basic Tools—Sources of High Energy Particles

Nature has given us two sources of fast particle flux: radio-activity and cosmic rays. Natural radioactive sources cannot furnish energies as high as those required for elementary particle physics — if radioactive substances had energies of more than about 10 *MeV* available, then their constituent neutrons and protons would be driven apart almost instantaneously.

Cosmic rays — the flux of particles from outer space hitting the earth — are of basic interest in themselves because they bring us important information about energetic processes in the universe. As sources of particles for high-energy physics studies their importance has decreased; this is a result of the increasing per-

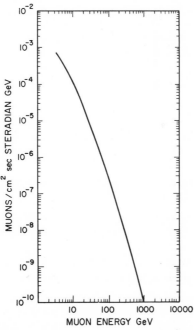

Figure 4-1. *Energy spectrum of cosmic-ray produced mu-mesons. Adapted from S. Bober* et al., *Nuclear Physics* B4, 539 (1968).

353

formance of particle accelerators. Until quite recently, almost all new particles were discovered through cosmic rays: the positron, the muon, the pion and the first "strange particles" were all discovered in cosmic-ray experiments. Even the anti-proton, although discovered by accelerator experiments, was recognised retroactively to have left its tracks in photographic plates exposed to the cosmic radiation.

The reason both for the early history and the present role of cosmic rays is clear: The flux of cosmic rays decreases very rapidly with energy; a typical energy spectrum of mu-mesons at the earth's surface is shown in *Figure 4-1*. If one considers that the intensity of the accelerator projected to furnish 200 *GeV* protons at Batavia, Illinois, in the U.S.A. is 10^{13} particles/sec, we can see that, as the energy of accelerators pushes upward, cosmic rays have a hard time competing. As one moves to higher elevations, or even into space, the flux of cosmic rays increases and therefore at the very highest energies (above 10^{12} electron volts or *TEV*) cosmic rays remain our only useful of particles, despite their low flux and uncontrolled direction. Even this situation will probably not persist as a new class of accelerating machines, called "colliding-beam storage rings" is perfected; we will come to this later.

All accelerators for high-energy particles use electric fields to act on the charge carried by the particle. In its most elementary form an accelerator is simply a pair of electrodes attached to a source of voltage V; a particle carrying an elementary charge e will gain an energy eV when traversing from one electrode to the other. Such an "honest" accelerator (see *Figure 4-2*) is limited fundamentally by the magnitude of the voltage which in practice can be placed between the two electrodes. Much technological progress has been made in this direction, and as a result the "electrostatic generator" has become an exceedingly useful tool in the energy range up to 20 *MeV* or so; it furnishes well-controlled, well-collimated and intense particle beams for nuclear structure physics, nuclear chemistry, studies of radiation effects on materials, and other technical applications. Electrostatic generators have been built in many forms. The voltage can be produced by direct transport of electric charge (e.g., using the belt of a Van de Graaf generator), by high-voltage transformers, or by charging condensers in parallel and discharging them in series.

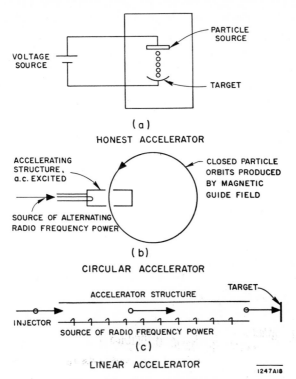

Figure 4-2. Type of accelerators.

The success of these "honest" producers of millions of electron volts is undeniable, but particle physicists are now interested in sources above the *GeV* range.   Since there is no practical means of placing billions of volts across electrodes, the larger accelerators of interest here must somehow elude this limitation.   As we will show, this can be achieved by the use of time-varying electric and magnetic fields.   Such accelerators can in turn be classified into two types: circular accelerators and linear accelerators.   In a circular accelerator the particle orbits are bent by a magnetic "guide field" into closed orbits so that the particles can traverse a region of accelerating electric field repeatedly.   In a linear accelerator, the particles are accelerated in a straight path (as in the electrostatic machines), but the electric fields are produced and controlled in such a way that they provide acceleration only in the regions where a "bunch" of particles happens to be. *Figure 4-2* indicates the basic principles of these types of accelerators in highly simplified form.

High-energy particle physics demands a variety of accelerators, not so much regarding the accelerating principles involved, but rather in terms of the characteristics of the beam produced. Among such beam properties are the following:

(a) *Kind* of beam particle (electrons or protons).

(b) Particle *energy*.

(c) *Intensity* or beam current (particles/sec).

(d) *Time structure* of beam (continuous or chopped into "bunches"— in the latter case the beam is "on" only for a fractional period called the "duty-cycle").

(e) *Beam geometry* (beam available inside the accelerator only, or as an "external" beam; collimation of beam, i.e., parallelism and size of beam spot; freedom from stray radiation around beam, etc.).

(f) *Energy spectrum* (the spread in energy of the particles in the accelerated beam).

All these properties are important in varying degrees in controlling the usefulness of the accelerator for experimental research; the types of experiments that are possible is largely determined by these accelerator characteristics, and it is this fact that creates the need for variety.

Figure 4-3 is a graph showing the world's accelerators operating about 1 *GeV*. The figure indicates three of the important parameters: kind of particle accelerated, energy, and intensity.

Let me now turn to a brief description of the principles of operation of each accelerator type.

The Cyclotron

The cyclotron is the "grand-daddy" of all circular accelerators. Although this machine in itself cannot produce particle beams above about 100 *MeV*, its basic principles are important here. We have seen in Chapter 3 (Eq. 3.7) that in a uniform magnetic field B charged particles would describe circular orbits of radius R given by

$$R = p/eB \qquad (4.1)$$

If the mass of the particle is a constant m_0, i.e., if the velocity of the particle is not yet close to the speed of light, then $p = m_0 v$, and the time to complete a circular orbit is given by

$$T = \frac{2\pi R}{v} = \frac{2\pi m_0}{eB} \qquad (4.2)$$

which is *independent* of the energy of the particle. This principle was recognised by E. O. Lawrence in 1929, and it forms the basis of the cyclotron.

In such a machine (see *Figure 4-4*) a set of D-shaped electrodes is introduced into the magnetic field; across these "D's" a high alternating voltage is applied whose period is just the period of revolution of the particles as described above. A source of charged nuclei (produced by stripping electrons from atoms in a gaseous discharge), usually protons, deuterons, or helium nuclei, is located at the centre. These nuclei are attracted into one of the "D's"

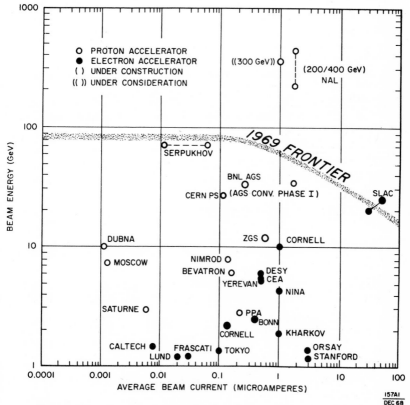

Figure 4-3. Comparative graph of various accelerator operating parameters, 1968.

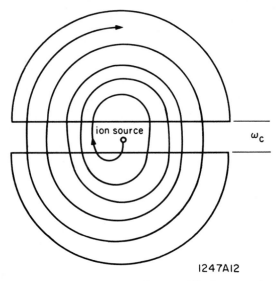

1247A12

Figure 4-4. Principle of the cyclotron. The ions are formed in the ion source and are drawn out by one of the D's. The particles are bent into circular orbits; the period of the orbit is equal to the period $(2\pi/\omega_c)$ of the applied D voltage. The particle will thus cross the gap between the D's at constant phase.

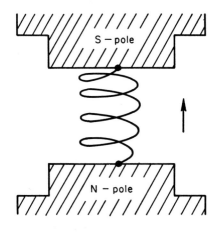

1247A9

Figure 4-5. Motion in a uniform magnetic field of induction B.

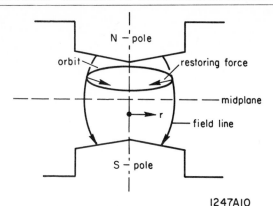

1247A1O

Figure 4-6. Motion in a magnetic field of circular symmetry with radial fall-off.

thereby gaining energy. They are then turned around by the magnetic field and cross the gap to the other "D". During the turn-around time the voltage across the electrodes has changed sign so that acceleration is produced again. This process continues until the particles spiral to the edge of the field.

Actually things are not quite as simple as all that. First, the orbits in a uniform magnetic field are not necessarily confined to a circle but a helix (spiral) is formed (see *Figure 4-5*). The reason is that the orbits can drift along the lines of magnetic field without any force to drive them back. This can be prevented by tapering the poles so that the field falls off with radius; as shown in *Figure 4-6,* this will produce a restoring force pushing the orbits back into the mid-plane of the magnet. But this substitutes one set of problems for another: if the magnetic field gets weaker as the radius increases, the period T (given by Eq. 4.2) will lengthen, and the particles will fall out of step.

This situation is further complicated by a second difficulty: relativity. You all know that the mass of a particle is in fact not constant as assumed, but increases with velocity; therefore the momentum p is not just simply given by $p = m_0 v$ as assumed above. The relativistically correct relation is

$$p = m_0 v / \sqrt{1 - v^2/c^2} \qquad (4.3)$$

and thus the period T becomes:

$$T = 2\pi m_0 / \left(\sqrt{1 - v^2/c^2}\; eB \right) = \frac{2\pi R}{v} \qquad (4.4)$$

Figure 4-7.   The 90-inch cyclotron at LRL, Livermore, California.

Therefore, as the particle spirals out, the period T will increase for two reasons: B decreases, and the factor $\sqrt{1 - v^2/c^2}$ also decreases.   In the cyclotron there is only one way to beat this game: brute force.   This is applied by putting as much voltage onto the "D's" as possible so that the final voltage is reached in relatively few turns; therefore the particles don't have enough time to fall out of step.   However, this voltage is limited by spark-over and by the available *rf* power, and the energy reached by the cyclotron is thus limited to values that are too low for high-energy physics.   *Figure 4-7* shows a photograph of the University of California 88-inch cyclotron.

Phase Stability

For a while the relativistic limit on the cyclotron seemed to be an insuperable barrier.   However, in 1943, V. I. Veksler in Russia and E. M. McMillan in the U.S. independently conceived a way to overcome this problem; this principle, called phase stability, applies both to circular and linear accelerators, although we will only treat the first case here.

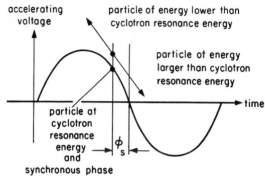

1247A13

Figure 4-8.   Principle of phase stability in circular accelerators. The diagram shows the accelerating voltage as a function of time. A particle of correct energy and radius in the magnetic field to be in cyclotron resonance will cross the accelerating gap at a phase angle Φ_s (synchronous phase) as shown; this phase is defined such that cos Φ_s is the ratio of energy gain per turn required by the accelerating programme divided by the maximum energy gain possible.   Particles of energy below resonance energy will tend to move toward phases of higher energy gain and vice versa; this action results in phase stability.

In *Figure 4-8* we plot the accelerating voltage across the "D's" of a cyclotron versus time.   Consider now a particle which crosses the gap between the "D's" during a decreasing part of this voltage wave.   A particle which has a little bit too much energy will describe a larger radius and will therefore be "late" at the next turn, thus receiving less energy.   A particle of energy less than "par" will get there early on the next turn and thus receive more acceleration.   The result is stabilising: the particles will oscillate stably about a specific operating radius and energy; this point is defined by the magnetic field and by the frequency of the *rf* oscillator.   If either the *rf* frequency, the magnetic field, or both, are varied in time the particle bunches will "ride" at the stable phase and the energy will adjust accordingly.   Therefore the "in-step" condition of the cyclotron will take care of itself.   However, one pays a price: Since the magnetic field and/or the frequency must change as the energy of the particles increases, the conditions at injection are different from those at full energy; as a result the beam emerges in bursts, rather than being continuous as in an ordinary cyclotron.

Accelerator	Frequency of Accelerating Voltage	Magnetic Guide Field	Radius	Time Cycle of Beam	Particle Accelerated
Ordinary Cyclotron	Constant	Constant in time; slightly decreasing with radius	Increasing	Continuous	Protons, deuteron, helium and heavier ions.
Synchrocyclotron*	Decreasing	Constant in time decreasing with radius	Increasing	About 1% duty cycle; about 60 pulses/sec.	Protons, deuteron and helium ions.
Electron Synchrotron	Constant	Time varying; focusing by radial decrease or strong focusing**	Constant	About 1% duty cycle; about 60 cycles/sec.	Electrons of velocity near that of light.
Proton Synchrotron	Increasing	Increasing focusing by radial decrease or strong focusing**	Constant	About 1% duty cycle; about 1 cycle/sec.	Protons.

* Called Synchrophasotron in Soviet literature.
** Strong focusing will be discussed later.

Figure 4-9. Table of circular accelerators; all but the ordinary cyclotron are phase-stable.

Figure 4-10. View of the vacuum tank and upper magnet coil of the 184-inch synchrocyclotron of the University of California Radiation Laboratory, Berkeley. Pumps are in the foreground; target-handling equipment is at the lower left and the rf oscillator housing is at the right-hand edge. Note the magnet return yoke and the shielding at the top of the picture. (University of California.)

The Synchrocyclotron

The phase-stability principle can be used in several ways. *Figure 4-9* gives a table showing the variety of applications. For lower energy (100-700 *MeV*) protons the synchrocyclotron is generally most practical; this machine looks like an ordinary cyclotron (see *Figure 4-10*), but the frequency of the accelerating system is cycled by using a rotating tuning condenser like those in an ordinary radio receiver (only bigger!). This varies the quantity T in Eq. 4.4, and, as the frequency goes down, T goes up and so does the energy of the particles. At the same time, of course, the radius increases and therefore the particles spiral out as in an ordinary cyclotron; however the voltage on the "D's" no longer needs to be high since operation is stable and many turns are permitted.

The Synchrotron

At still higher energies it becomes too expensive to build the huge electro-magnets in which the particles can travel from small to big radii. It is cheaper to build *ring magnets*, and to hold the particle orbits at constant radius. To do this while the particle energy increases, one cycles the magnetic field from a low to a high value. Particles are injected at low energy, are accelerated, and then are ejected to hit an external or internal target. This is the principle of the *synchrotron*, which is the basis of all of the existing circular accelerators operating above 1 *GeV*.

In practice synchrotrons can be designed in many ways. First, the requirements are very different depending upon whether electrons or protons are to be accelerated. Let us consider the electron synchrotron first. Since electrons are light (rest energy $0 \cdot 51$ *MeV*), they can be injected into the synchrotron at an energy already several times their rest energy, using an electrostatic accelerator (or a short linear accelerator — to be discussed later). As a result electrons already travel at almost the speed of light when entering the synchrotron ring. Therefore the accelerating frequency can remain constant, since both radius and velocity are constant, and the only quantity varying is the magnetic field.

There is one special problem with electron synchrotrons: light particles such as electrons will radiate light and X-rays when they travel in a circle at high energies. In fact a current of 1 *A* of

Figure 4-11. The 10-GeV electron synchrotron at Cornell University.

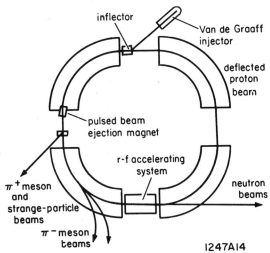

Figure 4-12. Schematic diagram of the principal components of a proton synchrotron. Note the various possibilities of external neutral beams, charged beams, and extracted primary beams.

3 *GeV* electrons bent into a circular loop by a 20,000 gauss magnet will radiate away energy at a rate of 1 megawatt!   The power radiated in a given field goes up as the square of the energy, so this is a serious limit.   This radiation loss hurts in two ways: The radio frequency accelerating system must be powerful enough to make up for these enormous energy losses, and the radiation hits the walls of the vacuum chamber enclosing the beam which therefore has to be adequately cooled and designed to withstand the beating.   Nevertheless, highly successful electron synchrotrons have been built up to an energy of 10 *GeV*; *Figure 4-11* shows a picture of the Cornell machine, the highest energy electron synchrotron in existence.

Protons of extremely high energy in strong magnetic fields do not radiate appreciable amounts of energy in the form of X-rays and light; therefore no limit other than cost presently exists to the energy at which a proton-synchrotron can be built.   Protons cannot be injected into the synchrotron ring from an electrostatic generator at velocities near that of light; therefore *both* the frequency of the accelerating system and the magnetic field have to be varied together so that the radius (as given by Eq. 4.4) will be constant. This usually requires a very wide swing in frequency.   *Figure 4-12* shows how the basic components of a proton synchrotron fit together, and *Figure 4-13* shows a photograph of the highest energy proton synchrotron now existing in the world, the 76 *GeV* machine at Serpukhov, U.S.S.R.

Figure 4-13.   The 70-GeV proton synchrotron at Serpukhov,
U.S.S.R.

Strong Focusing

The magnet of all synchrotrons must be designed to focus the particles so that they will be confined to the inside of the "doughnut"-shaped vacuum chamber in the magnet ring.   For the higher energy machines the magnet is expensive, and the cost is directly related to the volume of the magnetic field.   Hence there

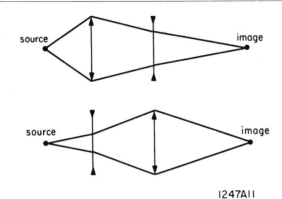

1247A11

Figure 4-14. Model of strong focusing: A combination of a converging and a diverging lens yields a net converging system.

is great interest in reducing the cross section of the doughnut; this in turn is only possible if the focusing action of the magnetic ring is very strong so as to squeeze the particles into tighter orbits.

The strength of the focusing action produced by tapering the magnetic field as shown in *Figure 4-6* has its limits: if the field taper becomes too large one can see by constructing the actual orbits that the particles spiral out of the field, i.e., the action becomes defocusing in the horizontal plane as the vertical focusing becomes stronger. This fact had been long recognised and appeared to produce a limit to the strength of focusing which could be produced. It was first discovered in 1950 by Nicholas Christofilos, and slightly later and independently by Courant, Livingston and Snyder, that this limit can be beaten.

Actually the principle, now called strong focusing, is quite simple; it had in fact been recognised earlier for other applications. It had been known for a long time that a combination of a pair of positive and negative lenses of equal focal lengths, when separated by a resonable amount, will produce a net focusing action (see *Figure 4-14*). Now we have just mentioned that a ring magnet with a *strong* fall-off of the magnetic field from its centre will be vertically focusing but horizontally defocusing. Conversely, a magnet with a radial increase of magnetic field will be radially focusing but vertically defocusing. If we therefore

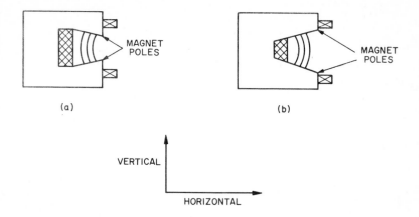

Figure 4-15. Magnetic element pair used in alternating-gradient focusing, also called strong focusing. The beam vacuum chamber passes through the region between the poles of each magnet in succession. Magnetic field lines are shown. In (a), beam particles are focused horizontally, but defocused vertically. The particles then go through (b), where they are defocused horizontally and focused vertically. The net result of the two-element magnet pair is focusing in both directions.

build our ring magnet not of uniform sectors, but of individual magnets which *alternate* in providing a decreasing or increasing magnetic field in the radius, then a net focusing action, both vertically and horizontally, results.

Figure 4-15 shows a cross section of magnets of the two types which constitute the magnets of the *alternating gradient synchrotron.* This type of focusing is now called "strong focusing", in contradistinction to the focusing action discussed earlier in *Figure 4-6* which is now called "weak focusing". The highest energy accelerators in the U.S., Western Europe and the U.S.S.R. (the Serpukhov accelerator is shown in *Figure 4-13*) are all alternating gradient machines.

The history of particle accelerators illustrates fully how necessity is the mother of invention. Whenever a limitation on the next step in energy appeared insuperable, or when costs for the next higher energy step appeared to be too great, new ideas removed the barriers. In the next lecture we will discuss how the obstacle toward highest energy for electron accelerators, as set by the emission of excessive radiation in circular electron accelerators, is removed.

CHAPTER 5
(W. K. H. Panofsky)

The Basic Tools—The Stanford Linear Accelerator

In Chapter 4 we examined the basic principles of circular accelerators from the cyclotron to the modern synchrotron. An important gap in the coverage possible by these circular machines is the lack of high-energy, high-intensity electron machines. This gap is closed by the electron linear accelerator.

Acceleration of particles in a long straight line requires high electric fields (i.e., a large electric force acting on the particle) but not necessarily high voltage from one end to the other. This can be accomplished by sending an electromagnetic wave which carries a component of electric fields along its direction of propagation, and by adjusting the speed of the wave to match the speed of the particle to be accelerated. This principle is illustrated in *Figure 5-1.* The particle appears to run "downhill" and therefore gains energy, yet since both the "hill" and the particle move ahead at the same rate, the particle actually does not change its height from the crests or the valley. This is the basic theory of the electron linear accelerator; now let us turn to more detailed questions.

An electromagnetic wave in free space (radio-waves, visible light, X-rays) travels with a fixed velocity ($c = 2 \cdot 998 \times 10^8 m/sec$); since the electric field direction associated with the wave is at

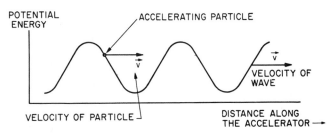

Figure 5-1. Linear acceleration of particles by electromagnetic wave propagation.

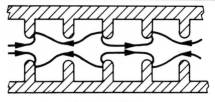

E FIELD LINES

Figure 5-2. Plot of lines of electric force in disc-loaded waveguide.

right angles to the direction of propagation, such a wave in free space is not useful for our purpose. If we now confine the wave inside a smooth metal pipe, the situation changes. The electric fields associated with the wave become more complicated, and in certain types of waves (called *TM* waves for "transverse magnetic" field) there is a longitudinal electric field which is suitable for acceleration. However, once a steady wave pattern in such a smooth pipe, called a waveguide, has been established, the wave actually propagates with a velocity greater than that of light in free space. Since, according to the theory of relativity, a particle always moves at a speed less than that of light, the particles and the wave in a smooth waveguide will necessarily fall out of step.

This problem can be solved by using what is called a "disc-loaded" or "corrugated" waveguide. In this case the velocity of the wave-pattern can be controlled by the geometrical configuration of the structure to almost any desired value, *Figure 5-2* illustrates the configuration that is actually used in Stanford's two-mile-long electron linear accelerator.

The frequency of oscillations of the electric field and the dimensions of the structure determine the speed of propagation of the wave; the relation is such that both the dimensions and the frequency must be held to very high accuracy. In the Stanford machine the frequency is 2856 *MHz* and the structure tolerances are of the order of $0 \cdot 0001$-inch; therefore, the temperature of the structure must be held to about $1/2°C$ despite the high power dissipated.

The reason why there is power dissipation in the walls of the structure is clear:. The oscillating electric field induces charges on the surfaces of the disc, and these charges flow back and forth along the outer wall. The resultant wall currents flow in a thin "skin" on the inside of the guide; this current-carrying layer is less than a thousandth of an inch thick; this so-called "skin depth" layer varies inversely as the square root of the frequency. As a result one can show that the energy gain ΔV in an accelerator section of length L, operating at a freuency f, when a power P is dissipated, is given by

$$\Delta V \text{ proportional to } \sqrt{PLf^{1/2}} \qquad (5.1)$$

Therefore, to obtain high-energy one requires both high radio-frequency power and large length, and the costs associated with each of these commodities should be approximately matched.   If one puts numbers into this equation, then one finds that for multi-*GeV* accelerators the power required becomes excessive unless one keeps it on only for a small fraction of the time, i.e., unless the accelerator is pulsed.

Let us illustrate this point by tabulating the parameters that are actually used in the Stanford two-mile accelerator (*Figure 5-3*). The total *rf* power required to produce 20 *GeV* in a length of two miles would be about 5000 megawatts at a frequency of 2856 *MHz*; considering the fact that high-frequency power sources generally operate at an efficiency of 50% or so, we find that 10,000 mega-watts of input power are required.   In the Stanford machine this is the peak power required, since the accelerator is pulsed 360 times per second at a pulse length of somewhat above 2 microsec. Thus the average power is $2 \times 360 \times 10^{-6} = 7\cdot2 \times 10^{-4}$ times smaller, or about 10 megawatts.   Of course, this is still a lot of power, but is generally less than the physicists use in carrying out experiments with the beam.

What are the best power sources for this application?   They should be high-gain amplifiers, rather than self-excited oscillators, because a common source of *rf* "driving power" is needed to tie the frequency and timing (phase) of each of the power sources together.   In the Stanford machine a new wave is started every 10 feet; four of these sections are driven from a single high-powered amplifier tube called a klystron.   Each klystron has a power amplification factor of 10^4, and each is driven by a low-power signal from a source feeding the entire machine. *Figure 5-4* is a schematic diagram showing how the klystrons are hooked up to the accelerating sections through wave guide "plumbing".   Both the waveguides connecting the klystrons to the accelerator and the

GENERAL ACCELERATOR SPECIFICATIONS

	STAGE I	STAGE II
Accelerator length	10,000 feet	10,000 feet
Length between feeds	10 feet	10 feet
Number of accelerator sections	960	960
Number of klyctrons	245	960
Peak power per klystron	6 – 24 MW	6 – 24 MW
Beam pulse repetition rate	1 – 360 pps	1 – 360 pps
RF pulse length	2.5 μsec	2.5 μsec
Filling time	0.83 μsec	0.83 μsec
Electron energy, unloaded	11.1 – 22.2 GeV	22.2 – 44.4 GeV
Electron energy, loaded	10 – 20 GeV	20 – 40 GeV
Electron peak beam current	25 – 50 mA	50 – 100 mA
Electron average beam current	15 – 30 μA	30 – 60 μA
Electron average beam power	0.15 – 0.6 MW	0.6 – 2.4 MW
Electron beam pulse length	0.01 – 2.1 μsec	0.01 – 2.1 μsec
Electron beam energy spread (max)	± 0.5%	± 0.5%
Positron Energy	7.4 – 14.8 GeV	14.8 – 29.6 GeV
Positron average beam current*	1.5 μA	1.5 μA
Multiple beam capability	3 interlaced beams with independently adjustable pulse length and current	
Operating frequency	2856 Mc/sec	2856 Mc/sec

* For 100 kW of incident electron beam power at positron source located at 1/3 point along accelerator length.

Figure 5-3. Parameters of the Stanford two-mile linear accelerator.

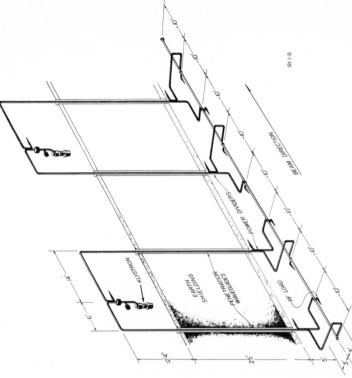

Figure 5-4. Schematic diagram of waveguide network connecting klystrons to accelerator sections.

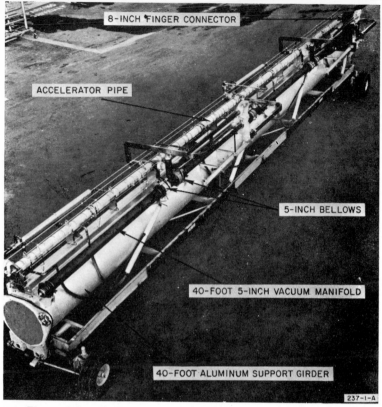

Figure 5-5. 40-foot accelerator subassembly showing principal parts.

Figure 5-6. Accelerator in underground housing.

accelerator itself have to be pumped down to a high vacuum since otherwise electrical breakdown would occur.

Since intense radiation is produced when even a small fraction of the beam strikes the accelerator wall, the disc-loaded waveguide structure and its support are placed under 25 feet of earth shielding. As a result, the feed-guides are rather long. Both the accelerating and the feed guides were manufactured in the Stanford Laboratories from a very pure (impurities less than 10 parts per million) *OFHC* (Oxygen-free High Conductivity) copper. This material was brazed in an atmosphere of hydrogen without the use of flux (which would introduce undesirable contamination into the vacuum system), and the dimensions of the components remained stable during the brazing operation. The accelerator components are precision machined, then pre-assembled in 40-foot lengths on a 2-foot diameter aluminum tube; such a subassembly is shown in *Figure 5-5*. These 40-foot units are then joined together in the underground tunnel to constitute the actual accelerator.

Figure 5-6 shows an inside view of the tunnel as it looks today. Because of the high levels of radiation inside the housing, most of the instruments such as valves, gauges, pumps, electronic controls, etc., are not placed inside the tunnel but are located above the shield in the "klystron gallery". Moreover, no organic gasket materials (which would be damaged by radiation) are used inside the tunnel; the accelerator pieces are joined by welding.

The klystron gallery (*Figure 5-7*) is a single building 30-feet wide, 14-feet high, and 10,000-feet long. In addition to the required instrumentation and the klystrons, the gallery contains the "modulators", which are the electrical power units which deliver the pulsed voltage required to feed the klystrons. The modulators store the electrical energy as it is being supplied continuously from DC rectifiers in banks of condensers; these condensers are then discharged in precise synchronism by a set of electronically driven switches, called thyratrons.

A difficult set of problems has to be solved to make these components work together as a "system". Let us mention but a few:

1. How do we align the accelerator to the needed accuracy (better than 1 *mm* in two miles) to get the beam through the holes without excessive loss?

2. How do we "phase" the different klystrons so that the waves in each 10-foot section are timed to smoothly hand the electron over from one wave crest to the next?

3. How do we "educate" the beam to pass through a set of holes only about 3/4-inch in diameter for a distance of two miles?

4. Finally, how do we know what the beam is doing so that the operators of the machine can make the necessary changes if the electrons do the wrong thing (such as not coming out of the far end at all!)?

Let us deal with these problems one at a time. First the problem of alignment: The individual 10-foot sections are placed on the 40-foot girders shown in *Figure 5-5* with high precision in the shop using small telescopes to line things up. The 40-foot pieces are then located to fair accuracy in the tunnel using the ordinary tools of the surveyor: transits, levelling telescopes, and rods. Finally once the pieces are joined together a laser beam is transmitted through the two-foot diameter pipe which supports the sections; a specially designed screen (also visible in *Figure 5-5*) can be flipped on command into the laser beam; this screen casts an image at the injection end of the machine which is electrically scanned; as the result one can determine how far each accelerator section is from a straight line to a few thousands of an inch.

Second the problem of phasing. The length of all waveguides, cables, etc., is first adjusted as well as can be done with electronic methods. Then the final adjustment is made by feedback using the beam itself. The operator can turn off each klystron one at a time. The "bunches" which ride the crest of the wave then induce a high-frequency signal in the "dead" sections which is electronically compared in timing (phase) with the klystron output of "live" sections. Any mis-timing can then be automatically adjusted. This process must be repeated whenever there is evidence that the energy of the beam is less than it should be or if components (such as the klystron tube itself) have been replaced.

Now how do we guide the beam through the small holes? First it turns out that the problem is not as hard as it seems, due to the acceleration process itself. Let us assume that after injection into the "regular" accelerator sections the beam diverges at an angle ϑ

Figure 5-7. Klystron gallery.

from the axis (*Figure 5-8*). We can describe this beam divergence by saying that the beam has a radial momentum p_r and a longitudinal momentum p_z, such that $\tan \vartheta = p_r/p_z$. After the electrons have passed further down the accelerator, the longitudinal momentum p_z has increased, while the radial momentum p_r has remained the same since the accelerator produces only negligible radial forces; therefore, the angle ϑ decreases. To say it more pictorially: the accelerating process "sucks" the particles into the tube and therefore the rate of spreading decreases. Note that the increase in longitudinal momentum takes place through increase in the *mass* of the electrons, not their velocity. By the time the electrons have been accelerated for a few feet they are moving essentially at the velocity of light. (The relativistically correct formula for the velocity v is $v = c \sqrt{1 - (m_0 c^2/E)^2}$; since $m_0 c^2 = 0 \cdot 51\ MeV$ the last term under the square root becomes negligibly small. For example, at an output energy of 20 *BeV* the speed of the electrons differs from c by only three parts in ten billion! On the other hand, the mass of the electron has increased 40,000-fold.)

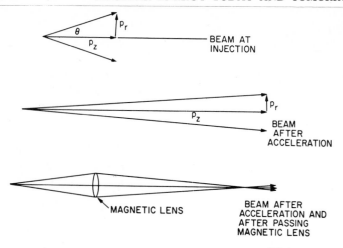

1247A16

Figure 5-8. Particle beam divergence at injection; decreased spreading with acceleration; and merger through strong-focusing.

The process of getting the beam through the machine is even further improved by the addition of strong-focusing (see Chapter 4) lenses every 300 feet along the machine. *Figure 5-8* shows how such lenses (shown in the photograph in *Figure 5-9*) bend the beam back to the axis. All this discussion shows that the initial aim of the particles at injection need not be as good as one might guess.

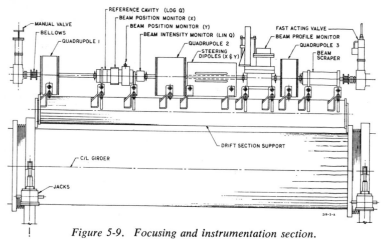

Figure 5-9. Focusing and instrumentation section.

There is also a fancier, more highbrow, way of looking at this situation: the special theory of relativity shows that if an observer rides with a speed close to that of light past a measuring stick, the length of this stick will appear shorter. As a result the length of the two-mile accelerator "appears" to the electron to be only two or three feet (!) while the size of the holes is unchanged. Therefore, the electrons, even without the presence of the lenses, have

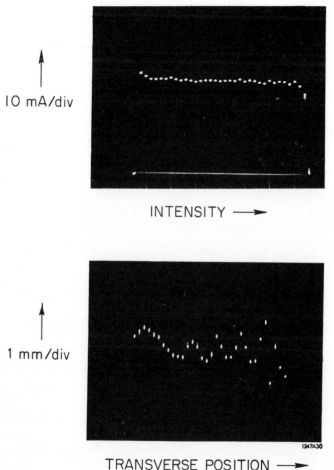

Figure 5-10. Oscilloscope display of beam intensity and position signals.

only to be aimed initially with sufficient accuracy to hit a hole of 3/4-inch diameter at a distance of a few feet, and that is not too hard a job.

Finally, how do we know what is happening to the beam as it tears through the two miles of accelerator in ten microseconds (10^{-5} sec)? We have to use instrumentation which itself does not disturb the beam. Such devices have been developed and are shown as little rectangular boxes in *Figure 5-9*. The beam in passing through these boxes (placed every 330 feet) excites electrical oscillations; the kind and strength of these oscillations measure both the intensity and transverse position of the beam. The resulting signals are processed electronically and are displayed on oscilloscopes to the operators. (See *Figure 5-10*.) If the beam is not well centred, the operator can adjust small electromagnets to "steer" the beam to its proper place.

We have described the essential components and system of the accelerator, but now we have to make the resultant beam useful to experimenters. Because of the high cost of the entire installation

Figure 5-11. Aerial photo of beam switchyard and research area.

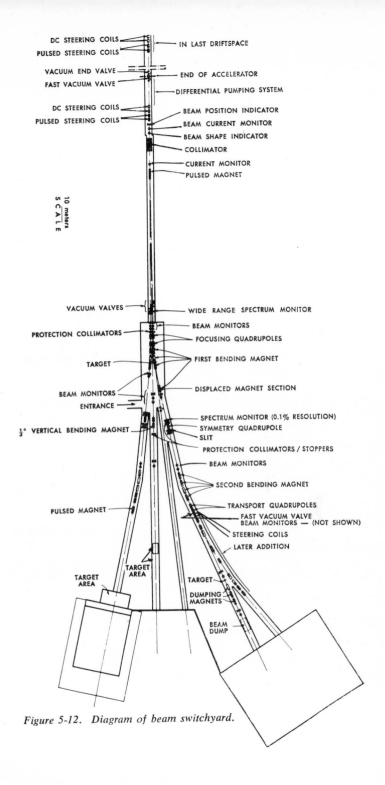

Figure 5-12. Diagram of beam switchyard.

Figure 5-13. Aerial photograph of SLAC.

and because of the large diversity of experiments physicists wish to perform, it is desirable to distribute the beams to several simultaneous users. This is done through a combination of magnetic focusing lenses and bending magnets assembled into what is called the beam-switchyard (BSY). *Figure 5-11* shows an aerial photograph of the BSY, and *Figure 5-12* is a diagram showing the location of the detection devices. We cannot describe here how all the individual devices function, but we can show what the overall system does. The BSY not only distributes the beam to the various physicist-customers, but it also precisely determines the beam energy at which the electrons can pass to the experimentalists. Moreover, the BSY can divide the beam pulses among different experimental set-ups because the first set of magnets is pulsed: i.e., the strength of their magnetism can be programmed to direct the beam either straight ahead, or to the left or right into different beam channels set to accept a specified energy. As a result of this and other instrumentation, each experimenter can receive any combination of pulses specified in energy, intensity, pulse length and pulse rate. Thus the 360 pulses produced by the accelerator each second can be shared so that each experimenter in essence has control over his own machine. This is particularly important since the different detection devices used by the physicists (described in Chapter 6) have a variety of needs for pulse rate and intensity. For instance, the bubble chamber generally cannot cycle faster than two pictures per second; therefore such a chamber uses only a small fraction of the pulses, and the rest can be given to experimenters using techniques which can operate at a high rate.

These are then the pieces which make up the accelerator, from the injector on the west end to the experimenters' locations in the target area toward the east. *Figure 5-13* shows an overview of the whole plant and gives some impression of its considerable size.

CHAPTER 6

(W. K. H. Panofsky)

The Basic Tools—Particle Detection and Analysis

In the previous two chapters we have studied the accelerators needed to produce high-energy particle beams; we now turn to the instruments designed to use them in particle research.

All detectors have to solve one basic problem: How to convert the minute amount of energy released when a single fast particle interacts with matter into a signal which can be recorded either photographically or electrically. From our discussions in Chapter 3 we know roughly how much energy we have to work with: a fast particle loses by ionization about 2 *MeV* of energy for each *gram/cm²* of matter it traverses. We can express this in joules by multiplying the loss in *eV* by the elementary unit of charge, $1 \cdot 6 \times 10^{-19}$ coulombs. Thus the energy loss is only $3 \cdot 2 \times 10^{-13}$ joules/(g/cm^2) or about $0 \cdot 7 \times 10^{-13}$ calories/(g/cm^2). Since this is not much energy to work with, the detector has to manage to amplify this energy or to cause it to *trigger* some larger scale unstable phenomenon. It is this latter situation which pertains to all "pictorial" detectors, i.e., those devices which photograph the tracks of particles.

The unstable phenomena most useful here are in the following categories:

1. Superheating of a liquid (the basis of the bubble chamber).
2. Supersaturation of vapor (the basis of the cloud chamber).
3. Supervoltage across electrodes (the basis of the spark and streamer chambers).

In applying these "super" phenomena to particle detection we proceed as follows. One arranges the conditions in the volume of the detector so that a large scale change of status of the material would be expected to occur; however, the change does *not* occur because the atoms in the material "do not know where to start". In fact, nothing happens until the passage of a charged particle with its minute energy loss triggers the beginning of the phenomenon along the track.

Figure 6-1.   SLAC 82-inch bubble chamber.

Let me illustrate this with the case of a superheated liquid.   If very pure water is heated in a smooth vessel its temperature will generally rise *above* the boiling point of 100°C (212°F); however, if the liquid is then disturbed, boiling will start explosively at the source of the disturbance.   We can produce the same result by suddenly reducing the pressure on a liquid. Since, as you know, water boils at a lower temperature at reduced pressure (such as that experienced at high mountain altitudes), forced expansion of a liquid can also cause superheating. This is the principle of the bubble chamber (BC): the BC is simply a container of liquid held at a temperature just below the boiling point. The pressure is then suddenly reduced just before passage of the particle to be studied. Boiling originates along the track of the particle, and lights are flashed to illuminate the bubbles which are then photographed. Many liquids can be used. The most popular is, of course, hydrogen because then interactions with elementary protons can be studied. However, heavy liquids such as freon and propane are also used in some chambers in order to increase the chances of making neutral secondary particles visible through their interactions in the liquid.

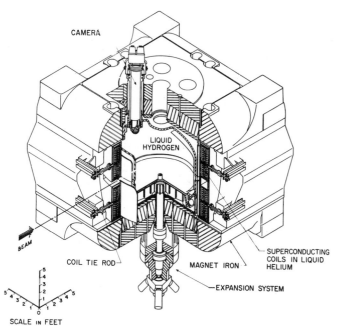

Figure 6-2. Argonne 12-foot bubble chamber.

Bubble chambers have many advantages. They can record the tracks of all charged particles in all directions, and very complicated processes can thus be seen directly. Low beam intensities are sufficient, and under well-controlled conditions the density of bubbles gives data on the degree of ionization. They also have disadvantages. Since they see everything they are unselective, i.e., observation of rare events requires analysis of many pictures that are not of interest. (Remember that the chamber has to be expanded before it is known whether anything of interest has occurred.)

As we will see later the hydrogen bubble chamber (HBC) has been the most important single tool for the discovery of new phenomena in elementary particle physics. *Figure 6-1* and *Figure 6-2* show photographs of modern chambers which are housed in large electromagnets to permit momentum measurements. *Figure 6-3* is a photograph taken with a beam of charged particles ($K^{\pm}$ particles) showing some interesting events, and *Figure 6-4* is a picture taken in a high-energy γ-ray beam. Note that the incident track in *Figure 6-4* is not visible (γ-rays do not

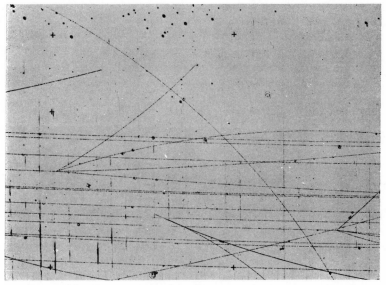

Figure 6-3.   Events produced by K^+ mesons in the SLAC 82-inch bubble chamber.

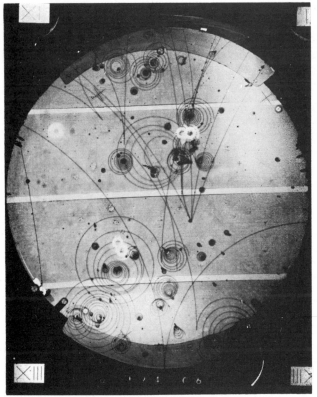

*Figure 6-4.   Events produced by high-energy gamma rays in the
SLAC 40-inch bubble chamber.*

ionize) and that many low-energy electron-positron pairs are
produced.

The cloud chamber is no longer used in high-energy physics,
but historically it was responsible for many discoveries, e.g., the
positive electron and the mu-meson were discovered in cloud
chambers. In the cloud chamber a supersaturated state is produced by sudden expansion of a mixture of water vapor and air;
when a particle passes through the chamber, it begins to "rain"
along the track and the drops can be photographed.

Another "visual" detector which has passed the peak of its
usefulness for high-energy particle physics (using accelerators) is
nuclear emulsion. By "nuclear" emulsions we mean special

photographic emulsions of much greater than normal thickness (up to 1 *mm* compared to a few microns) which have a larger than usual content of silver salts.   When a fast particle passes through such an emulsion it leaves a latent image along its track which can be rendered visible by photographic developers.   Such emulsions can be sensitive enough to record tracks at the minimum

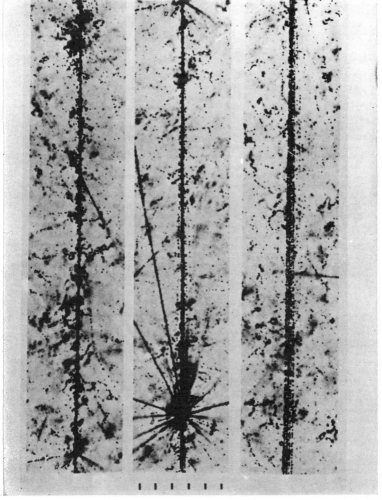

Figure 6-5.   Cosmic ray induced event in nuclear emulsion.

of the curve of energy loss versus particle velocity. Currently the principal use of emulsions is for cosmic-ray experiments since they can be easily carried aloft by balloons or rockets, and require no auxiliary equipment. Historically they also played a significant role around accelerators: The first artificially produced π-and μ-mesons were detected in emulsions by Gardner and Lattes in 1948. *Figure 6-5* shows an ultra-high-energy cosmic-ray event as seen in an emulsion detector.

Let us now look at a more recent pictorial detector, the optical spark chamber. Its principle is old, but its successful application is recent. When a high voltage is applied to electrodes in a gas atmosphere, once the voltage exceeds a certain limit, an electric breakdown, or spark, will occur. Air, for instance, will break down if a voltage gradient in excess of 30,000 volts/*cm* is applied. If an overvoltage is applied across electrodes suddenly, electrical breakdown will occur only after some delay unless a disturbance triggers it. Such a disturbance is provided by the

Figure 6-6. A multiplate spark chamber with particle tracks.

free charges liberated by an ionizing, i.e., charged, particle, and therefore a spark will occur where such a particle has crossed the region across which the large electric field has been applied. *Figure 6-6* shows a multiplate spark chamber made of aluminum plates across which a high voltage has been placed suddenly, followed by the passage of a fast particle.

The spark chamber has a great advantage over the bubble chamber: it is selective. The bubble chamber has to be expanded before there can be any knowledge whether an interesting event has occurred. On the other hand the ions formed by a charged particle track "hang around" long enough so that the high voltage across a spark chamber can be applied *after* passage of the particle. Hence some auxiliary set of electronic particle detectors (like the scintillation counter to be discussed later) can be sensitized to particularly interesting events; the decision that such an event has occurred can be made electronically and the voltage is then suddenly applied. Therefore spark chambers are used in conjunction with other detectors which identify or define a class of interesting events, with observation of the details of the event being left to the spark chamber.

The sparks in the chamber give a good deal of light; photography is thus easy. On the other hand, the spark will not accurately follow the path of a particle across a large gap, so that precise track measurements are difficult. Moreover, it is more difficult to use the spark chamber if many particles pass through it because the electrical energy tends to divide unevenly among the sparks. However, spark chambers are easy to build and are a flexible tool in that the plate material, shape and thickness can be adapted to the needs of any particular experiment. For these reasons various spark chamber arrangements are used for a large fraction of current experiments in high-energy physics.

The deficiencies of the spark chamber can be circumvented if the voltage across the plates is not only applied suddenly, but is also turned off again after a duration of only a few nanoseconds (1 nanosecond = 10^{-9} sec). In that case a full spark does not have time to form; rather, there is a faint discharge, called a *streamer*, which appears along the particle tracks. The light output of these streamers is so small that photography requires

Figure 6-7. Large streamer chamber (in magnet) at SLAC.

extreme sensitivity. The streamer chamber combines certain
advantages of both the bubble chamber and the spark chamber. It
can record in all directions and it also permits accurate measurements of curved tracks in a magnetic field; at the same time it
can be selective by being "triggered" in response to identification

Figure 6-8. *Events produced by a high-energy gamma ray beam in a large streamer chamber.*

Figure 6-9. *Wire spark chamber.*

of interesting events. Its main disadvantage relative to the bubble chamber is that the medium in which the tracks are formed is a gas rather than a liquid, and that hydrogen cannot be used as the "streamer" medium. Thus the interaction rates are lower, and if reactions in hydrogen (i.e., on the proton) are to be studied, then a separate "bag" of hydrogen must be introduced (in which case the point of origin of an event cannot be seen). *Figure 6-7* shows a picture of the large streamer chamber at SLAC housed in its magnet. *Figure 6-8* illustrates a typical photographed event.

All the pictorial methods described thus far end up with tracks recorded photographically. This is where the largest job begins: the scanning of the film for important events, the measurement of the curvature of the tracks and of the angles between them, and then the reconstruction of the dynamics of the process studied. All this could not be done without massive use of modern computers. Although the initial recognition of an event is usually done by eye, the information contained on the photograph is then transcribed into magnetic computer tape either manually or on special measuring tables, or automatically by various devices. This is a gigantic task. In the U.S. alone, over 10 million such pictures are processed per year.

It is this giant effort to digest the information contained in track chamber photographs which has led physicists to search for means to by-pass the need for photography and to invent means to couple a spark chamber electronically to the computer. There are several ways to do this. One is acoustic. When a spark occurs it produces a sharp "crack" which can be picked up by microphones. From the different times of arrival of the sound at several microphones one can locate the position of the spark; therefore, using the electrical output from the microphones the computer can locate each spark and the entire track.

This method somewhat lags in accuracy behind the optical chambers, and has been largely replaced by another device called the wire spark chamber (*Figure 6-9*). Here the electrodes of the spark chamber are not flat sheets but are grids of parallel fine wires; once a spark jumps, a current will flow in just one of the wires. Ingenious methods have been devised to read out "which wire has the current"; the simplest is to have each wire loop through a magnet core similar to the kind used in computers. The

6-10. Oscilloscope display of wire spark chamber output.

magnetic cores are magnetised by the current in the particular wire to which the spark has jumped, and the computer examines the state of magnetisation of each core. Obviously it takes several such wire planes to locate the spark exactly, but actual chambers are quite inexpensive to build. An obvious gain is that the physicist can get some of the results of the experiment as he goes along. *Figure 6-10* shows a photograph of an oscilloscope screen on which the computer has reconstructed an event seen by a group of spark chamber planes.

The wire chamber is a sophisticated outgrowth of the pictorial detector; it is no longer "pictorial" but is really a collection of individual particle detectors producing electronic signals. It therefore belongs to a different family of detectors, that of "electronic counters", the earliest and best known of which is the Geiger counter. Geiger counters are gas-filled tubes incorporating two electrodes across which a high voltage is maintained; when a fast particle passes through the gas, a discharge occurs.

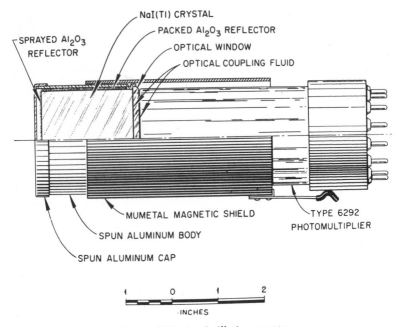

Figure 6-11. A scintillation counter.

Geiger counters are no longer used as detectors in high-energy physics; their place has largely been taken by other devices, notably the scintillation counter, an example of which is shown in *Figure 6-11.* The principle is the following: When a charged particle penetrates matter the electric impulse caused by the passage near an atom can knock an electron out of the atom's shell; this process, as we discussed earlier, is called ionisation. However, in many cases the electron is not knocked out but is only displaced from its ordinary orbit around the nucleus to a new orbit which has more energy; this process is called "excitation" of the atom. After some time the electron will return to its original orbit; the extra energy available is radiated in the form of light. As the result many transparent materials "scintillate" when traversed by a charged particle. The most popular scintillators used today are plastic. They are made of polystyrene plastic loaded with certain organic chemicals. The light output is viewed by a "photo-multiplier" tube which is a photoelectric device with built-in internal amplification; the electric output from this tube is then processed and recorded.

The modern scintillation counter is really a rebirth of an old device; Rutherford in his α-particle scattering experiments in 1911 used scintillation caused by the impact of the particles on a scintillating ZnS screen to record their presence. Of course he used eyes rather than photomultipliers as light detectors!

In high-energy physics experiments scintillation counters are rarely used singly; in general they are used in combination to map out how deeply particles penetrate into an absorber, or how they are distributed in space after being bent by a magnet, and so on. The light pulse from a scintillation counter is very fast; therefore such counters can be used to measure the time-of-flight, and hence the speed of particles even if they travel close to the speed of light.

Finally, I would like to mention the "Cerenkov" counter. We discussed in Chapter 3 that a particle can travel with a speed which is faster than the speed of light in a transparent medium (but not faster than the speed of light in a vacuum!); in that case a light flash is produced which propagates along the surface of a cone (see *Figure 3-8*), similar to the shockwave produced in air by a supersonic airplane. This phenomenon can be used to advantage as a particle detector. Again a photomultiplier can be used to record the light flash. The Cerenkov counter can be used in a much more selective manner than a scintillation counter. Since the condition for having Cerenkov radiation is that $v > c/n$ (where v is the particle velocity, c is the velocity of light in free space, and n is the refractive index), we can make the counter selective to particles of different speeds by varying the refractive index n. This can be done by filling the counter with high pressure gas; the quantity $n - 1$ is generally proportional to gas pressure and therefore we can adjust the minimum value of the velocity of a particle which can be detected. Such a high pressure Cerenkov counter is shown in *Figure 6-12*.

In this chapter we have surveyed the arsenal of particle detectors available to the high-energy experimentalists, ranging from pictorial devices such as the bubble chamber to specialised particle counters.

These in combination with bending and focusing magnets, radiation shielding and electronic processing and recording devices make up the experiments which have given us the "inside story" of the particles in the nucleus.

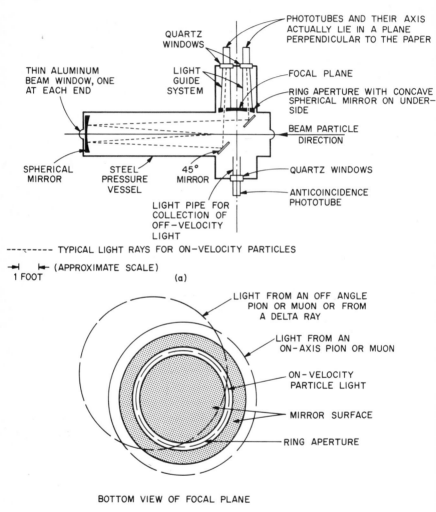

Figure 6-12. (a) Schematic diagram of Cerenkov counter optics. (b) Illustration of focal-plane behaviour of the Cerenkov light-ring images from different particles.

CHAPTER 7
(R. H. Dalitz)

Symmetries and Selection Rules

We have seen that much of our experience with systems of elementary particles and their reaction processes can be summarised in the expression of several selection rules. We now wish to review all of these empirical rules and to discuss their theoretical basis.

All selection rules can be regarded as being a reflection of some symmetry property possessed by the underlying physical laws, for example by the equations of motion or by the interaction energy. Some of these symmetry properties appear natural a priori, some of them can be given an elegant and aesthetically pleasing form but were quite unexpected and far from apparent a priori, and others again are purely ad hoc symmetries, devised to lead to the selection rules known empirically.

The symmetries associated with the properties of space and time appear rather natural. The *conservation of momentum* follows directly from the assumption that space is uniform, i.e., that the laws of physics remain of the same form when our reference frame is translated without rotation. The *conservation of energy* is associated with the assumption of uniformity in time, i.e., that the laws of physics include no reference to any particular origin in time. The *conservation of angular momentum J* is a direct consequence of the assumption that three-dimensional space is isotropic, i.e., that the laws of physics retain the same form when the reference axes are rotated to a new orientation. This last statement must be extended somewhat, to take account of the intrinsic spins for all the particles of the system. The intrinsic spin (which may include internal orbital angular momentum) is a quantized quantity, whose values are limited to $S\hbar$, where $\hbar$ is is Planck's constant, and S is limited to integral or half-integral values. Then, for a system of particles labelled $i = 1, 2, 3 \ldots n$, conservation of the total angular momentum

$$J = \Sigma_i \, (L_i + S_i) \tag{7.1}$$

follows from the assumption that the laws of physics retain the same form when all of the reference axes, including the reference axes for all the spin variables, are given the same rotation to a new orientation.

The other space-time symmetries possible are with respect to space reflection in a point (the operation P), and with respect to the reversal of the time axis (the operation T). P is generally known as the parity operation. When this operation is carried out twice in succession, the axes are back to their original configuration, so that we have $P^2 = 1$. When the laws of physics are space reflection invariant (or P-invariant), we can take a given state and combine it with the space-reflected state to make states which are even or odd with respect to P. We call this sign the *parity* of the state, even parity for $P = +1$ and odd parity for $P = -1$. If P-invariance holds for the strong and electromagnetic interactions, then each "elementary particle" state will have a definite intrinsic parity which can be determined

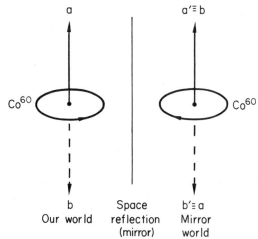

Figure 7-1. *The left figure represents schematically a Co^{60} nucleus spinning with right-hand motion about the upward axis a, and we are considering the situation where the electron emitted from the beta-decay of this nucleus travels along the axis a. This situation is reflected in a mirror, and then appears as in the right figure, where the reflected Co^{60} now spins left-handed about the upward axis. a' is the reflection of a, so the emitted electron moves along this direction a'. However, emission in the direction a' in the mirror world is equivalent (in so far as the laws of physics are concerned) with emission along direction b for Co^{60} in our world.*

empirically (we shall give two examples below). The parity of the state of a system of particles is the product of the intrinsic parities together with definite parity values characteristic of the orbital motions of the particles. If the interaction energy is P-invariant, then the total parity must remain unchanged through any reaction processes the system may undergo; in other words, *parity conservation* must then hold.

We emphasise here that the important point is the P-invariance of the laws of physics appropriate to a given physical situation. This can be best illustrated by an example, given in *Figure 7-1*. Before discussing this, we should mention that it is more convenient in practice to consider mirror reflection rather than reflection in a point. These two operations differ only by a $180°$ rotation about the axis normal to the plane of the mirror; since we have already accepted that the laws of physics are invariant for a rotation of axes, the two operations are essentially equivalent in their consequences. Let us return to *Figure 7-1*; the figure on the left represents schematically the nucleus Co^{60} spinning in a right-handed sense about the axis marked a. We consider the intensity n_+ of the electrons emitted along a when the Co^{60} nucleus undergoes beta-decay. Now consider the same physical system in the mirror-reflected axes. In this mirror world, the Co^{60} nucleus spins with a left-handed sense about the axis directed along a'; however, the intensity of the electrons emitted along a' is again n_+, since these are actually the same emission events, but seen in the mirror-reflected world. If the laws of physics are the same in the mirror world, then they must say that the intensity of electrons emitted along b' (since b' is the axis about which the Co^{60} spins with a right-handed sense) is n_+. Hence parity conservation (P-invariance) for all the laws of physics involved in these beta-decays requires that the intensity of emitted electrons must be the same along either direction a' or b'. Any up-down asymmetry in the beta emission from Co^{60} polarised with a definite sense of spin about a vertical axis would imply that parity conservation does not hold for some interaction involved in this decay process.

Symmetry of the physical laws with respect to time reflection does not lead to any conservation law. Rather, this symmetry implies a definite relationship between a given reaction process $AB \rightarrow CD$ and inverse reaction process $C^T D^T \rightarrow A^T B^T$, where the superfix T denotes that the momentum and the spin of the particle are to be reversed.

One further symmetry operation which we have already emphasized to be intimately linked with the structure of special relativity, although it is not strictly a space-time operation, is the particle-antiparticle conjugation (or charge reflection) C. This operation simply replaces every particle by its antiparticle, leaving spins and momenta unaffected. Hence, C also leads to a relationship between particle reaction processes and antiparticle reaction processes, rather than to a conservation law, since it reverses the quantum numbers B, Q and s. The only exceptions to this remark are the systems for which $B = Q = s = 0$, such that C transforms the system into itself. Examples of this situation are provided by a photon, a π^0 meson, or an η-meson. C-invariance requires $C = -1$ for the photon (since C reverses the charge and current to which it is coupled), and the observed decay processes $\pi^0 \to \gamma\gamma$ and $\eta \to \gamma\gamma$ require $C = +1$ for π^0 and η mesons.

From a cosmological viewpoint, it is possible to query the invariance assumptions underlying the above conservation laws or symmetry relationships. We do live in a particular part of the universe, and perhaps it is not completely valid to ignore this fact. Space is not isotropic here; we live in a galaxy which is shaped more like a pill-box than a uniform sphere. There may be an origin of time, associated with the early history of our universe. However, we are concerned here with the subnuclear world involving distances and time-intervals which are as remotely small from our every-day experience as the galactic phenomena are remotely large. It appears reasonable to assume that the properties of space on the subnuclear scale should be independent of phenomena occuring on the galactic scale, and this has been the working hypothesis generally adopted in elementary particle physics today. No evidence has yet appeared to the contrary.

Although it is easy to devise theories with interactions which violate P-invariance, C-invariance, or T-invariance (or any two of them simultaneously), it has not been found possible to construct any theory whose interactions are not invariant with respect to the three operations C, P and T taken together. This CPT-invariance has very great generality. It does have physical consequences; for example, it is *CPT*-invariance which requires that the partial lifetime for the process $A \rightarrow abc$ . . should be equal to the partial lifetime for the antiparticle process $\overline{A} \rightarrow \overline{abc} \ldots$, and this has been tested empirically to considerable accuracy for a number of processes. We should note that, assuming *CPT*-invariance, *T*- invariance has the same consequences as invariance with respect to $(CPT)T = CP$ (since $TT = 1$). Logically, how-ever, the operations *T* and *CP* are quite distinct and there is a good deal of experimental work going on to test the two operations separately, especially for the weak decay interactions.

For the strong interactions, all the evidence is that *P*-, *C*-, and *T*-invariance all hold good, at least to 1% accuracy or better. For electromagnetic phenomena the situation is rather less clear at present, especially as far as *C*-invariance or *T*-invariance are concerned, and there are experiments going on in many laboratories to test these invariance properties severely; a typical experiment is the comparison of the rates for the two photo-processes

$$\gamma + D \rightleftharpoons N + P \qquad (7.2)$$

On the other hand, the evidence that *P*-invariance holds good for electromagnetic processes is quite strong.

We now give several examples to illustrate how these conserva-tion laws or selection rules are used to determine the quantum numbers of individual particles. Consider the charged π-meson. Its spin *S* has been determined most directly by counting the number $(2S + 1)$ of spin states (with $S_3 = -S, -S + 1, \ldots + S$) it has. This is clearly involved in the comparison of the reaction rates for the following two processes

$$\pi^+ + D \rightleftharpoons P + P \qquad (7.3)$$

The total rate for the forward reaction is for a single pion, averaged over all its spin states, and therefore has a factor $1/(2S + 1)$. The total rate for the backward reaction is for all final spin states of the pion, and does not have this factor. The experimental comparison between these rates indicates $S = 0$ quite definitely. The $\pi^{\pm}$ parity is determined from the observation that the reaction

$$\pi^- + D \rightarrow N + N \qquad (7.4)$$

occurs for a π^- meson coming to rest in deuterium. A very convincing argument indicates that the π^- capture occurs after the meson reaches the lowest level in the π^--D atom. An analysis of the final state then shows that angular momentum conservation (the only angular momentum in the initial state is $J = 1$ for the deuteron) requires negative parity for the final system. Hence, we conclude spin-parity $(JP) = (0 -)$ for the π^+ mesons.

For the π^0 meson, assuming spin $S = 0$, the parity can be determined from the relation between the polarisation planes for the two photons from its decay $\pi^0 \to \gamma\gamma$. The two planes are predicted to be parallel for $P = + 1$, and perpendicular for $P = - 1$. This determination has actually been carried out from observations on the two $e^+ - e^-$ planes in the double internal pair conversion process

$$\pi^0 \to (e^+ + e^-) + (e^+ + e^-), \qquad (7.5)$$

leading again to the conclusion of negative parity for the pions.

The study of the way in which the energy released in the decay processes $K^+ \to \pi^+\pi^+\pi^-$ and $\eta \to \pi^+\pi^-\pi^0$ is shared among the pions leads to the conclusion that these pions carry no orbital angular momentum, so that the final parity just arises from the intrinsic parities of the pions. For the η-meson, the conclusion is that the spin-parity is $(0 -)$, as for the pions. The same conclusion also holds for the final three pions in the K^+ decay (the τ-mode) and indicates spin zero and negative parity. However, the decay process $K^+ \to \pi^+\pi^0$ also occurs (known as the θ mode); here again there is no orbital angular momentum (since all the spins are zero) but the final intrinsic parities now give total parity positive. The discrepancy between these two conclusions was known as the τ-θ puzzle, and provided the first indication that the assumption of parity conservation might not hold valid for the weak decay interactions, as we shall discuss in a later lecture.

The second class of conservation laws are those involving the additive quantum numbers, as follows:

(i) charge Q,
(ii) baryon number B,
(iii) electron number N_e,
(iv) muon number $N\mu$.

As far as we know, these conservation laws hold exactly. This is true, to an extraordinarily high degree of accuracy, for charge conservation and baryon conservation. For the lepton numbers N_e and N_μ, we know only that they hold to quite a good accuracy (perhaps to 1% in rate), but our present ideas about the nature of the weak interaction (to be discussed in a later lecture) would require the conservation of N_e and of N_μ to be separately exact.

We can ensure that these conservation laws hold for the appropriate interactions by demanding that these interactions have an appropriate symmetry property. For the case of charge, this symmetry is already familiar to us from classical electrodynamics. It is the property of *gauge invariance*, which expresses the fact that the electrodynamic forces which act on a charged particle depends only on the field strengths E and H. No physical differences arise when the electromagnetic potentials A and V are changed in such a way that the field strengths remain the same. The notion of a gauge field is a very deep one. For Q, it provides an answer to the question "how can two charges at different space-time points be compared?": if this is not possible then how could we assert the conservation of charge. The electromagnetic field provides us with a means to determine the charge on a particle at any point by measuring the force which this field exerts on the particle, in other words to gauge the charge on the given particle. There has been a good deal of controversy about the mass appropriate to the quantum of this gauge field; I think the purist would insist that, insofar as the gauge invariance is exact, this mass should be zero, as is indeed the case for the photon. With an exact gauge invariance, the property of charge conservation then follows necessarily.

Many attempts have been made to extend this notion of a gauge field to the other additively-conserved quantities, especially for B and s. The main difficulty is simple; there exists no other zero-mass quanta analogous to the photon, as would be required by the existence of gauge fields for B and s. However, it is still possible to lay down a simple set of rules for the construction of theories which will necessarily obey the conservation law considered, but these rules are then ad hoc, derived from the constraints observed in the experimental phenomena and giving no depth of understanding why this conservation law should hold.

The third class of conservation laws are only approximate. This means that the corresponding symmetry can hold exactly only for certain parts of the interaction energy, while other parts with weaker couplings do not have this symmetry. So far we have met two examples of this class, the strangeness s and the isospin I associated with the occurrence of hadronic charge multiplets.

Strangeness conservation holds rather exactly for all nuclear and electromagnetic processes, but is violated by some of the weak decay interactions. This is not surprising, since the motivation for its introduction was to provide a means for distinguishing between these two classes of interaction. Attempts have been made to relate s-conservation with the existence of a gauge field, but no appropriate particle with zero mass is known.

The property of charge independence for the strong interactions is equivalent to an isospin symmetry for this part of the total energy. This symmetry is generally referred to as $SU(2)$ symmetry for the strong interactions; this means that all the members of a charge multiplet occur in an equivalent way in this interaction energy, which must remain invariant in form when a series of related transformations are made on all charge multiplets, mixing the members of each individual multiplet in a well-defined way. For the case of nuclei, the description $SU(2)$ corresponds to the fact that the basic objects are the nucleon doublet, and the basic transformation is essentially an arbitrary mixing of the neutron and proton states, for a given state of motion. When this symmetry is extended to other particles, such as the mesons and the baryons, this direct interpretation is no longer relevant, although the same transformations are still appropriate for any charge multiplet with a given isospin I, now interpreted only in an abstract and mathematical sense.

The $SU(2)$ isospin symmetry holds rather well for the strong interactions, but it is violated both by the electromagnetic interactions, since these depend on the individual charge for each member of the multiplet, and by the weak interactions, which generally lead to transitions (e.g., $K^+ \rightarrow \pi^+ \pi^0$) necessarily involving a change of isospin. Hence the isospin selection rules are expected to hold only to an accuracy of order 1% ($e^2/\hbar c = 1/137$), owing to the effects of the electrodynamic corrections.

CHAPTER 8
(R. H. Dalitz)

Hadron Physics and the Unitary Symmetry

The hadrons may be divided first into sub-classes, according to their *baryon number B*, with names as follows:

Baryon No. $B =$ 0	Mesons		B
$B = -1$	antibaryons	baryons	$+ 1$
$B = -2$	anti-dibaryons	dibaryons	$+ 2$
$B = -3$	etc.	etc.	$+ 3$

The only stable dibaryon we know is the deuteron D. Its antiparticle $\overline{D}$, the bound state of $\overline{P}$ and $\overline{N}$, has been observed in experiments at the Brookhaven 30 *GeV* proton accelerator. This table can clearly be extended arbitrarily in B, but our interest lies primarily with the simplest systems, those with $B = 0$ and $B = \pm 1$.

Next, we can classify the hadrons with baryon number B according to their spin-parity values. In fact, the semi-stable baryons mostly have spin-parity $(\frac{1}{2} +)$, except for the Ω^- particle which has $(3/2 +)$, and the semi-stable mesons are all $(0 -)$. There are also more massive particles which are unstable, but which all follow the same rule, that baryonic states have half-integral spin and meson states have integral spin. At this point, we must refer to an important result of great generality, known as the *Spin-Statistics Theorem*. This states that, for particles A with half-integral spin, only one particle of type A can exist in a given state of spin and orbital motion (this situation leads to the case of Fermi-Dirac statistics, appropriate to nucleons in a nuclear state, for example), whereas for particles B with integral spin, any number of particles B can exist in such a state (this situation leads to the case of Bose-Einstein statistics, appropriate to photons, for example). This theorem applies just as well to the leptons and the photon as to the hadrons.

Let us consider the $(\frac{1}{2}\,+)$ baryon states in Table I. We have already recognised that they occur in charge multiplets, the mass deviations within each multiplet being typically of order 5 *MeV*, attributable to the effects of the electromagnetic interactions, which violate the $SU(2)$ isospin symmetry. Including the multiplets with non-zero strangeness, the charge of each individual particle state is given by

$$Q = I_3 + (B + s)/2. \qquad (8.1)$$

It is convenient to give a special name to the quantity $Y = (s + B)$, the *hypercharge*, since its use leads to a greater symmetry in our representation for the baryons and mesons. With this, we then have

$$Q = I_3 + Y/2. \qquad (8.2)$$

When each baryon is represented by a point on the plane (I_3, Y), as shown on *Figure 2-3*, they form a hexagonal array, symmetrically placed about the origin, the two neutral particles being represented by points at the origin. It is of interest to note (see *Figure 8-1*) that the antibaryons (which have spin $(\frac{1}{2}\,-)$, the parity reversal being an effect of relativity which we cannot discuss in detail here) form precisely the same array when plotted on this

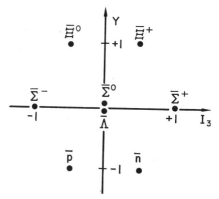

Figure 8-1. The octet of anti-baryons are plotted on a (I_3, Y) plane, each anti-baryon being represented by a point in the appropriate location. We note that this octet array is identical with the octet array for the baryons themselves, except that the labelling of the points is different.

plane. The (0 −) mesons, the π triplet, the K and $\bar{K}$ doublets, and the η singlet also form precisely this same pattern on the (I_3, Y) plane, as shown on *Figure 2-4*.

This octet pattern consists in each case of a *group of charge multiplets* with the same spin-parity and with comparable mass values. The existence of this pattern runs parallel to the existence of charge multiplets themselves, and our interpretation of it represents simply an extension of the notion of charge independence to *"charge and hypercharge independence"*. The eight particles in each octet are then considered to be essentially the same particle, at least with respect to some higher class of strong interactions, far stronger than ordinary nuclear forces. These "superstrong interactions", as we shall call them, do not recognise charge or hypercharge. The main difference between this situation and the situation for charge multiplets which we have discussed previously is that the mass differences between the particles in the same "charge and hypercharge multiplet" are much greater, typically $\delta M \approx 200 \, MeV$ for the baryons and $\delta M \approx 400 \, MeV$ for the mesons. This is a reasonable situation only if the characteristic energy associated with these superstrong forces is measured in thousands of *MeV*; otherwise it would be most unreasonable to have such strong traces of this higher octet symmetry surviving in the observed properties of these particles. The hypothesis that this *octet symmetry* may hold for the "elementary particles" was first put forward as the "Eightfold Way" by Gell-Mann in 1961, and independently by Ne'eman in the same year.

It is interesting to consider several direct tests for the accuracy with which this octet symmetry holds. These involve electromagnetic effects, and we must first recognise explicitly that the electromagnetic interaction is with the charge and current within the particle. Octet symmetry requires that the eight particles have precisely the same internal structure as far as the superstrong forces which dominate are concerned; in particular, two octet particles with the same total charge necessarily have the same internal currents and hence the same electromagnetic properties.

(i) First, we consider the contribution of the electromagnetic self-interactions within each particle to the mass of that particle. It is these electromagnetic self-masses which must be responsible

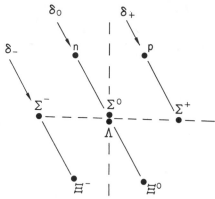

Figure 8-2. In this octet array for the baryons, lines are drawn connecting the baryon states which have the same charge. Since the baryons are all alike in structure, the baryons with the same charge have the same electro-magnetic properties. Here we are considering the electro-magnetic self mass δ, which takes three distinct values for the three distinct charge values occurring.

for the mass differences observed between particles within the same charge multiplet, if we believe that the strong nuclear interactions are charge-independent. Although we are not yet able to make quantitative calculations for these electromagnetic self-masses individually, it is possible to deduce a relationship (first given by Coleman and Glashow in 1961) between the mass differences within the various charge multiplets belonging to the same octet. The situation is illustrated in *Figure 8-2*. The octet states with the same charge $(+\ e,\ 0$ or $-\ e)$ lie on three lines across this pattern, making $120°$ with the I_3 axis; the self-masses associated with the particles on each of these three lines are δ_+, δ_0 and δ_-, respectively. Without electromagnetic effects, the neutron and proton would have precisely the same mass; hence their mass difference is given by

$$M(N) - M(P) = \delta_0 - \delta_+ \tag{8.3a}$$

Similarly, for the Σ and Ξ charge multiplets, we have

$$M(\Sigma^-) - M(\Sigma^+) = \delta_- - \delta_+, \tag{8.3b}$$

$$M(\Xi^-) - M(\Xi^0) = \delta_- - \delta_0. \tag{8.3c}$$

Subtracting (8.3b) from (8.3a) and equating this with (8.3c) gives the Coleman-Glashow relation,

$$M(\Xi^-) - M(\Xi^0) = M(\Sigma^-) - M(\Sigma^+) - (M(N) - M(P)) \tag{8.4}$$

The best values for the baryon masses give the value $6·63 \pm 0·17$

MeV for the right-hand side; the Ξ mass difference is not so accurately known but the empirical value is at present $6 \cdot 55 \pm 0 \cdot 7$ *MeV*, in agreement with the value predicted by Eq. (8.4), to better than 10% accuracy.

(ii)   the baryon magnetic moments are all predicted in terms of $\mu(P)$ and $\mu(N)$, the magnetic moments of the nucleons, whose values are

$$\mu(P) = 2 \cdot 7928 \text{ n.m.}, \ \mu(N) = -1 \cdot 9132 \text{ n.m.} \qquad (8 \cdot 5)$$

where the unit n.m. devotes the nuclear magneton $e\hbar/2M(P)c = 3 \cdot 15 \times 10^{-18}$ *MeV* gauss^{-1}. From *Figure 8-2* it is clear that Σ^+ and P must have the same electromagnetic properties, so that the octet model predicts $\mu(\Sigma^+) = \mu(P)$; the measured value for $\mu(\Sigma^+)$ is $2 \cdot 5 \pm 0 \cdot 5$ n.m., in excellent agreement with the prediction. The prediction that $\mu(\Lambda) = \mu(N)/2$ is less obvious; the measured value is $\mu(\Lambda) = -0 \cdot 73 \pm 0 \cdot 16$ n.m., not in disagreement with this prediction.

Gell-Mann also put forward a definite hypothesis about the character of the strong interactions which do not have this octet symmetry, namely that they correspond to the $I = 0$, $Y = 0$ component of an octet quantity. From this hypothesis it was possible to derive a general formula,

$$M(I, Y) = M_o + aY + b(I(I + 1) - \tfrac{1}{4}Y^2). \qquad (8 \cdot 6)$$

for the masses of the individual charge multiplets in a given charge-hypercharge multiplet, in terms of three constants characteristic of this multiplet. For the baryon octet, the elimination of these three constants leads to the mass relation

$$(3 M(\Lambda) + M(\Sigma))/4 = (M(\mathcal{N}) + M(\Xi))/2, \qquad (8 \cdot 7)$$

where $\mathcal{N}$ denotes the nucleon doublet. Using the mean masses for each charge multiplet, the left and right-hand sides of Eq. (8.7) have the values $1135 \cdot 0$ *MeV* and $1128 \cdot 5$ *MeV*, respectively, in good qualitative accord with this relation.

We note that the octet pattern of *Figure 2-3* has three axes with the same symmetry as that seen with respect to the I_3 and Y axes, the other two sets of axes being inclined at 60° and 120° to the present axes. This three-fold symmetry is due to the fact that the octet symmetry is intimately linked with an abstract $SU(3)$ symmetry (in fact, it is a special case of $SU(3)$ symmetry). This $SU(3)$ symmetry (which we will refer to frequently as *unitary symmetry*) represents a generalisation of the $SU(2)$ isospin symmetry, for which there are two additive quantum numbers I_3 and Y where the isospin symmetry has only the one, I_3; this was essentially the implication of the term "charge and hypercharge independence" which we used earlier to characterise these unitary multiplets of charge multiplets.

The $(0 -)$ mesons also form an octet pattern, to a good approximation, as was shown on *Figure 2-4*. The mass relation (8.6) is usually written for (mass)^2 in the case of mesons (this might well be more appropriate for the baryon case too); also the fact that particle and antiparticle must have the same mass but have opposite values for Y requires the absence of the term a. Thus, for mesons, we have the mass relation

$$(3 \ M^2(\eta) + M^2(\pi))/4 = M^2(K). \qquad (8.8)$$

Inserting the π and η masses on the left-hand side of this relation would lead to the prediction $M(K) = 480 \ MeV$, quite comparable with the mean K-meson mass, $495 \ MeV$, so that this relation also works quite well here.

With $SU(3)$ symmetry, we may well ask "is there a basic charge- and hypercharge-triplet on which these algebraic transformations act?", by analogy with the situation for $SU(2)$ symmetry in nuclear physics, where the basic doublet consists of the nucleons (N, P). Although no such triplet particles have yet been discovered, it has been convenient and remarkably fruitful to think about the hadronic particles as if they were little molecules or nuclei constructed from a basic triplet of particles and their antiparticles. The name *quark* has come to be used rather generally for these hypothetical triplet objects, their antiparticles being termed the *antiquarks*. It turns out that the simplest way in which a baryon octet can be constructed from triplet quarks requires that three quarks must be bound together to form a baryon. This naturally requires that the quarks should have baryon number $B = \frac{1}{3}$. This requires that they must have fractional values for their charge and hypercharge. They must also have half-integral spin, in order that the three-quark system can have spin $\frac{1}{2}$ in the baryon states; naturally we suppose that the quark spin is $S = \frac{1}{2}$.

The quark and antiquark states are shown pictorially on *Figure 8-3*, the doublet quarks with zero strangeness being denoted by

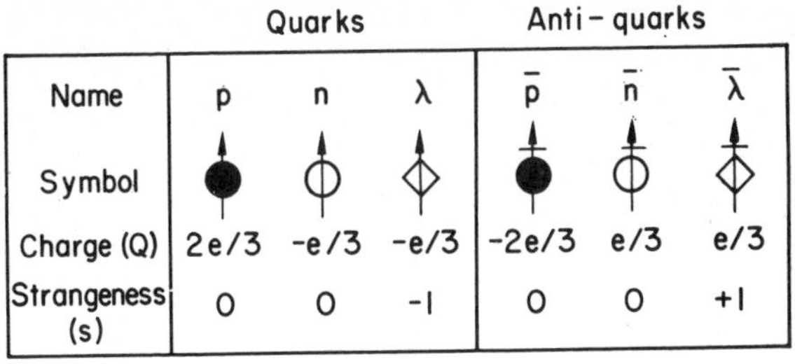

	Quarks			Anti-quarks		
Name	p	n	λ	$\bar{p}$	$\bar{n}$	$\bar{\lambda}$
Symbol	●	⊕	◇	●	⊕	◇
Charge (Q)	2e/3	-e/3	-e/3	-2e/3	e/3	e/3
Strangeness (s)	0	0	-1	0	0	+1

Figure 8-3. The quark triplet states are characterized by an appropriate symbol. The difference between p and n quarks is that the p quark has positive charge. The (p,n) quarks are non-strange, whereas the λ quark has strangeness s = − 1, and has therefore been denoted by a different symbol. The antiquarks are distinguished by a bar over the top of the symbol. Each antiquark has the opposite charge and strangeness to the corresponding quark.

(*p, n*) and the singlet quark with strangeness $s = - 1$ by λ. We note that the *p* quark has charge $+ 2e/3$, and the *n* and λ quarks have charge $- e/3$. These fractional values for the quark charges are not pleasing to the physicist. We have been accustomed to *e* as the unit of charge and it has been regarded as one of the outstanding puzzles in physics to understand why charge should be quantized, with just this unit value, and why this charge value should hold for such widely different particles as the proton and the electron. Now we are asked to accept the quark as the most fundamental object in strong interaction physics, but with charge values $\pm 2e/3, \pm e/3$! For this reason, many other schemes with basic triplet objects have been put forward, such as to allow integral charge values for all the triplet objects, but these schemes generally involve hypothesizing several distinct sets of triplet objects and thus become quite complicated in detail. The scheme based on a single quark triplet, with the fractional charge values given on *Figure 8-3*, is the simplest scheme possible, and we shall see that this already has sufficient complexity to account for all the "elementary particle" states reported to date. The outstanding difficulty about the quarks is that no such particles have yet been observed in Nature, in the free state, although their distinctive charge values should allow them to be detected with relative ease. We shall discuss the quark hunts which have been carried out already in a later lecture. We must emphasise that the existence of a quark triplet is not essential for the validity of $SU(3)$ symmetry, nor is it required by $SU(3)$ symmetry in any sense. If we accept the hypothesis of a quark triplet, then it is natural to suppose that the binding forces between them will be the super-strong forces, whose strength is large relative to the strong forces which violate the $SU(3)$ symmetry and which are measured by an energy typically 200 *MeV*.

In terms of these hypothetical quarks, the meson states result from the combination of a quark and an antiquark, with their spins antiparallel to give total spin $S = 0$. These states are represented pictorially in *Figure 8-4*, being arranged in columns according to the value of their charge Q and in rows according to their strangeness number s. We retrieve the symmetry shown on *Figure 2-4* by sliding the $s = 1$ doublet states appropriately to the right, and the $s = -1$ doublet states appropriately to the left, which corresponds to arranging the rows according to the isospin component I_3 rather than the charge $Q = I_3 + s/2$. We note that there are nine states formed rather than eight. Comparing these with the octet pattern of states on *Figure 2-4*, we see that the additional state corresponds to the centre point, with $Q = 0$, $s = 0$, or equivalently $I_3 = 0$, $Y = 0$. With any kind of symmetry scheme, it is always possible to have an isolated state (called a singlet state) which transforms into itself for every symmetry

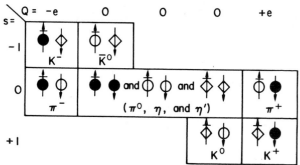

Figure 8-4. All the combinations of quark and antiquark which can occur together are tabulated, in rows according to their strangeness, and in columns according to their charge. The arrows denote their spin directions; the configurations considered here are all with opposing spins, therefore for total spin S = 0. There are altogether nine states, consisting of the (0 −) octet and the η′ unitary singlet with s = Q = 0, which is a combination of the three configurations given in this location.

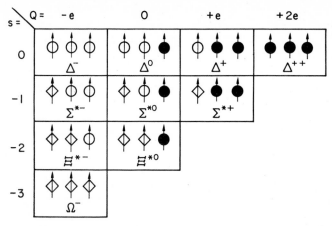

Figure 8-5. The configurations of three quarks which give rise to the baryonic decuplet, the individual states being arranged in columns according to charge Q and in rows according to strangeness s. The spins of the three quarks are shown parallel, so that they add to given total spin S = 3/2, as is appropriate for the physically observed decuplet states.

operation. With $SU(3)$ symmetry, a singlet state can only have the quantum numbers $I_3 = 0$, $Y = 0$. Hence, our belief is that the nine quark-antiquark states on *Figure 8-4* consist of an octet of (0 −) states, together with a singlet (0 −) state whose mass could be completely different from the mean mass typical of the octet states. In fact, there does exist such a (0 −) singlet state, the η′ meson at 958 *MeV*, which is not semi-stable but decays relatively slowly (lifetime believed to be about 10^{-21} sec), dominantly to the final state ($\eta \pi^+ \pi^-$). This η′ meson is most probably the ninth (0 −) meson represented as a quark-antiquark system in *Figure 8-4*.

The quantum numbers for the quarks have been chosen such that the baryon octet states are three-quark systems. It is then natural to ask the question in reverse, what other $B = 1$ $SU(3)$ multiplets could be formed from the combination of three quarks. The answer is that the only other possibilities are a singlet state (neutral and with strangeness − 1) and a decuplet array, which is laid out in *Figure 8-5* where the columns give the charge Q and the rows give the strangeness s.

CHAPTER 9
(R. H. Dalitz)

The New Spectroscopy

The distinction we have made between semi-stable particles and unstable particles is one of great practical importance for the experimenter who wishes to study them in detail. There is clearly a great difference between the study of the $\pi^{\pm}$ and $K^{\pm}$ mesons whose lifetimes are about 10^{-8} sec and for which rather pure beams with well-defined momenta may be constructed to allow a detailed study of their reaction, scattering, and decay processes, and the study of particles whose lifetimes are of the order of $T = 10^{-23}$ sec, and which can therefore travel only a distance of order $cT = (3 \times 10^{-10}) \, 10^{-23} = 3 \times 10^{-13} \, cm$, a distance of the order of the dimensions of a nucleus, between their formation and their decay. Nevertheless, this distinction has no fundamental significance, as we have already learned from our experience with the excited states of nuclei. We have already given one example of this in *Figure 2-2*, where the resonance state Be^6 was shown to be a member of a charge triplet, the other members being a $(0 +)$ excited state Li^{6*} (which decays by γ-ray emission) and He^6 (a semi-stable state which undergoes beta-decay to the Li^6 ground state). Whether the decay of a particular particle involves nuclear, radiative or weak decay processes depends, in a sense, on accidents of mass relationships, on the relation of its mass value to the energy thresholds for the many final states allowed for this particle decay by all the selection rules other than energy conservation. It should therefore have come as no surprise to elementary particle physicists that there should exist an enormous number of unstable particle states, since this situation was well-known for the atomic nuclei, but the fact is that physicists were so sure that the semi-stable particles they were investigating were really "elementary" that it took many years before this viewpoint was generally accepted, that these exceedingly short-lived objects were recognised to represent simply an extension of our table of "elementary particles" (for example, our Table I) upwards in mass, into mass regions where there always exist final states to which the particle can decay rapidly through the strong nuclear interactions.

The first unstable particle was discovered when $\pi^\pm$ beams became available for the study of their scattering by target protons (as the nuclei of the hydrogen atoms in a target containing liquid hydrogen). The quantity most simply measured is the total interaction cross-sections σ_{total} which is effectively the area presented to the incident pions by each target proton, which the pion must strike in order to interact with the proton; it measures essentially the probability for the interaction of the incident particle with the target particle, at a given energy. The cross-sections measured for the π^+P and π^-P

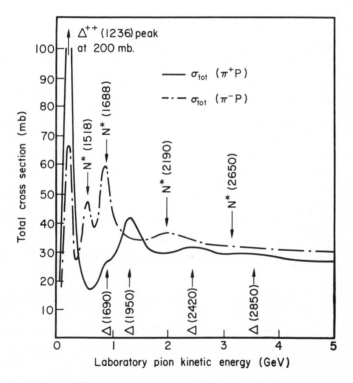

Figure 9-1. The total cross-sections for π^+P and π^-P interactions are plotted as function of the laboratory pion kinetic energy. The arrows indicate particle states, which are labelled N^* when they are charge doublet, and Δ when they are charge quadruplet.

interactions are shown on *Figure 9-1* up to laboratory pion kinetic energy of 5 *GeV*. At high energy, the cross-sections vary rather smoothly with increasing energy, but there are a number of striking peaks in these data for kinetic energies below about 2 *GeV*. Each of these peaks is interpreted as the excitation of a particle state B^*,

$$\pi + P \rightarrow B^*. \tag{9.1}$$

followed by its rapid decay back to the initial state (the process of elastic scattering) or to other states if there are any allowable with conservation of energy, thus, for example

$$B^* \rightarrow P + \pi, \tag{9.2a}$$
$$\rightarrow P + \pi + \pi \tag{9.2b}$$
$$\rightarrow \Lambda + K. \tag{9.2c}$$

We should note that the mass value quoted for each peak is the energy M^* of the particle B^* at rest. Its relation with the incident laboratory kinetic energy K is obtained simply by considering the invariant $(E^2 - c^2P^2)$ in the laboratory reference frame and then in the B^* rest frame, thus

$$(mc^2 + K + Mc^2)^2 - c^2P^2 = (M + m)^2c^4 + 2(M + m)\,K$$
$$= E^2 - c^2P^2 = (M^*c^2)^2 \tag{9.3}$$

where m and M denote the rest-mass for the incident and target particles; we have used the fact that the kinetic energy of the incident particle is given by

$$K = \sqrt{(m^2c^4 + P^2c^2)} - mc^2 \tag{9.4}$$

Since the particle B^* decays rapidly, with mean lifetime T, its mass value is correspondingly poorly defined. If Γ (*MeV*) denotes the mass spread (the width) at half-intensity on the cross-section peak, the mean lifetime T is given by

$$T = \hbar/\Gamma$$
$$= (6 \cdot 58 \times 10^{-22}/\Gamma(MeV))\ \text{sec.} \tag{9.5}$$

The most striking peak is that at $M^* = 1236\ MeV$, which occurs in both the π^+p and π^-p cross-section curves. It has width Γ about 110 *MeV*, which corresponds to a lifetime of only 6×10^{-24} sec. This particle Δ is a charge quadruplet which corresponds to isospin $I = \frac{3}{2}$. The states $\triangle^{++}$ and $\triangle^+$ are shown on Fig. 9-1, and there must also exist states Δ^- and Δ^0 which are the mirror particles to them (in the sense in which N is said to be the mirror particle to P, π^- is the mirror particle to π^+, and H^3 is the mirror nucleus to He^3; more properly, this mirror relationship is called charge symmetry and relates the two particles with $\pm I_3$ in the same charge multiplet). Its spin and parity are determined from the study of the angular distribution for the decay (9.2a); in this case, the distribution observed for the angle Θ between the directions of the outgoing pions are the incident pion fits rather well the form $(1 + 3cos^2\Theta)$, which is characteristic of spin 3/2 for the B^* particle. The parity is most readily shown to be $(+)$ by observations of the angular distribution of pions scattered from a polarised proton target, that is a target in which it is arranged that the spins of all the protons point in the same direction.

The peak at 1518 MeV in the π^-p cross-section is unmatched by any companion peak in the π^+p cross-section. This particle is a charge doublet, and can therefore appear only in states of charge 0 and $+$ 1. The angular distribution here indicates spin 3/2 but the polarisation experiments indicate parity $(-)$. The other prominent resonances are the charge doublet N^* (1688) with spin-parity (5/2 $+$) and Δ (1950) with spin-parity (7/2 $+$).

These resonances may also be excited by photons, thus

$$\gamma + P \to B^{*+}, \tag{9.6}$$

followed by B^* decay through modes such as those given in (9.2). In fact, the study of these photoproduction processes played a rather significant role in the story of the discovery of Δ (1236), and in the determination of the detailed properties of N^* (1518).

There are also a very considerable number of other nucleonic particles with mass values in the range covered by the cross-section data shown in *Figure 9-1*. Some are not prominent because they are excited only with difficulty through πP interaction; others are not seen because they are hidden under other prominent resonance peaks. Again, when there are many particles lying with mass values not widely separated, their separate resonance peaks may overlap strongly to give a net total cross-section curve which is quite smooth, giving little hint of the individual cross-section peaks beneath. This is probably the situation for the $\sigma(\pi P)$ curves above about 2·5 GeV pion kinetic energy, since we expect the number of resonance states to increase very rapidly with increasing mass and there are already at least ten nucleonic resonances between $M^* = 1$ GeV and $M = 2$ GeV.

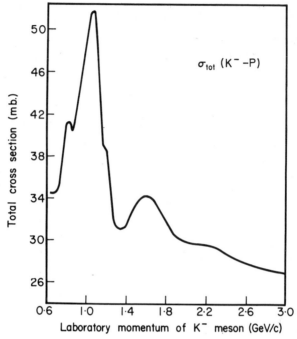

Figure 9-2. The total cross-section for the K^- – P interaction is plotted as function of the laboratory momentum of the K^- meson for the interval 0.6 to 3.0 GeV/c. The dominant peak is due to the charge singlet state Λ^* (1815), and shows two subsidiary peaks on its wings. The peak at 1.6 GeV/c arises from at least two distinct particle states.

Similar resonance peaks corresponding to unstable baryons also occur for other strangeness values. The states with $s = -1$ are explored in the $\overline{K^-P}$ interaction, for which the total cross-section is plotted against laboratory K^- momentum in *Figure 9-2*. This curve shows a striking peak at $M^* = 1815\ MeV$, which has been established to correspond to a charge singlet particle with spin parity (5/2 +). The mass range covered by the data shown in this figure includes at least ten resonances, several of which can be seen on the wings of Λ^* (1815), with others for laboratory momenta about $1\cdot6\ GeV/c$ and $2\cdot2\ GeV/c$. In contrast, the K^+P cross-section (not shown here) shows relatively little structure as function of momentum, apart from a rise from about 13 mb to 19 mb over the M^* interval 1700 to 1900 MeV, which is associated with the onset of inelastic pion-production processes. There is some controversy at present about the possibility of a resonance peak in the K^+N system, but this system is at present studied only for the K^+N interactions taking place in K^+D collisions, which complicates the interpretation of the actual experimental observations.

Not all of the higher particle states can be reached in such convenient experiments. The first $s = -1$ resonance to be established was the baryon Σ^* (1385), whose mass lies below the K^-P threshold $(M_P c^2 + m_K c^2 = 1432\ MeV)$. This particle can be identified only by examining its decay products. Its dominant decay mode is

$$\Sigma^* (1385) \rightarrow \Lambda + \pi, \tag{9.7}$$

and it was first identified in the reaction processes

$$K^- + P \begin{array}{c} \nearrow \Sigma^{*+} + \pi^- \searrow \\ \longrightarrow \\ \searrow \Sigma^{*-} + \pi^+ \nearrow \end{array} \Lambda + \pi^+ + \pi^- \qquad \begin{array}{c} (9.8a) \\ (9.8b) \\ (9.8c) \end{array}$$

Those $\Lambda\pi^+$ pairs which resulted from Σ^{*+} are characterised by the fact that their total energy $M(\Lambda\pi)$, measured in the reference frame where their total momentum is zero, is equal to the mass $M(\Sigma^*)$. The expression for $M(\Lambda\pi)$ in terms of the observed energies and momenta is obtained by considering the invariant $(E^2 - c^2 P^2)$ for this pair of particles, thus

$$M^2(\Lambda\pi)c^4 = (E_\Lambda + E_\pi)^2 - c^2 (\mathbf{p}_\Lambda + \mathbf{p}_\pi)^2. \tag{9.9}$$

The first state for each event (9.8) is completely specified by giving the values $M^2(\Lambda\pi^+)$ and $M^2(\Lambda\pi^-)$, and each event can therefore be plotted as a point on a graph whose co-ordinates are $(M^2(\Lambda\pi^-), M^2(\Lambda\pi^+))$. This has been done on *Figure 9-3* for about 1500 $\Lambda\pi^+\pi^-$ events obtained in an experiment using K^- mesons with 1510 MeV/c momentum from the Bevatron at Berkeley. We notice first that the events are concentrated very strongly along two bands, corresponding to events where either $M(\Lambda\pi^+)$ or $M(\Lambda\pi^-)$ has value in the range 1360-1400 MeV. These are the events of types (9.8a) and (9.8c), resulting from a two-stage reaction, and they establish the existence of the charge triplet particle $\Sigma^*(1385)$, which has width $\Gamma = 40\ MeV$. A detailed study of the characteristics of the decay process $\Sigma^* \rightarrow \Lambda_\pi$

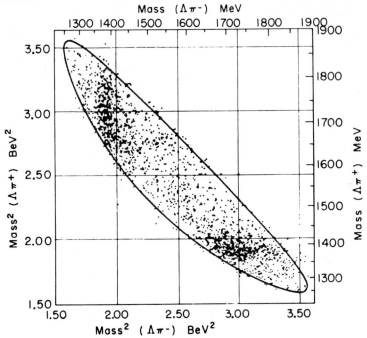

Figure 9-3. A plot of 1582 examples of the reaction $K^- + P \rightarrow \Lambda + \pi^+ + \pi^-$ for K^- laboratory momentum 1510 MeV/c, as observed by the Alvarez bubble chamber group at Berkeley, the abscissa being (mass)² for $\Lambda\pi^-$, and the ordinate (mass)² for $\Lambda\pi^+$. Strong resonance bands are seen with respect to both $\Lambda\pi^+$ and $\Lambda\pi^-$ for mass values around 1385 MeV, corresponding to the production and decay of $\Sigma^*(1385)^\pm$ in this reaction.

in these events, including information about the spin polarization of the Λ particle, have led to the spin-parity assignment $(3/2 +)$.

The baryons with strangeness $s = -2$ and greater can only be studied in this indirect way, through the various reactions such as

$$K^- + P \rightarrow K + \Xi^* \rightarrow K + (\Xi + \pi). \qquad (9.10)$$

The Ξ^* particle which has been particularly well studied is the low-lying charge doublet $\Xi^*(1530)$, for which the spin-parity assignment $(3/2 +)$ has been established. A spin-parity determination of $(5/2 +)$ has also been obtained recently for the $\Xi^*(2030)$ particle, another charge doublet. Several other Ξ^* charge doublets are definitely established but not very thoroughly studied yet.

One baryon with $s = -3$, the Ω^- baryon at 1672 MeV, has become well-established through observations on the production reaction

$$K^- + P \rightarrow K^0 + K^+ + \Omega^-. \tag{9.11}$$

It is a semi-stable particle, with lifetime about 10^{-10} sec and the decay modes $\Xi \pi$ and ΛK^-. Its stability results from the fact that its mass lies below the threshold $(M(\Xi)c^2 + m(K)c^2) = 1810$ MeV for the lightest two-particle system with $B = 1$ and $s = -3$.

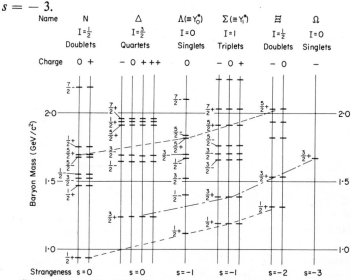

Figure 9-4. All of the baryon resonances established to date up to mass about 2100 MeV are plotted on this figure, in columns according to their charge, isospin, and strangeness. The dashed lines connect together states which definitely belong to the same SU(3) multiplet. The ordinate gives the total mass in unit GeV.

All of baryon states known up to mass about 2100 MeV have been plotted on *Figure 9-4*. The spectrum of levels is quite complicated. The charge multiplets which have the same spin-parity value and are known to belong to a definite unitary multiplet are linked by dashed lines. We have already discussed the $(\frac{1}{2} +)$ octet; now an octet of $(5/2 +)$ states has also been established. The $\Lambda^*(1405)$ and $\Lambda^*(1520)$ particles, with spin-parity $(\frac{1}{2} -)$ and $(3/2 -)$ respectively, appear to be well separated from any possible partner states with the same spin-parity value, and they are clearly unitary singlets.

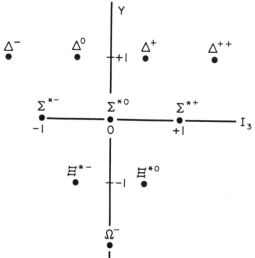

Figure 9-5. All of the baryonic states belonging to an SU(3) decuplet are plotted on a (I_3, Y) plane. Notice the three-fold symmetry of this array, which is not so apparent in the array of Figure 8-5 where the co-ordinates are (charge, strangeness).

The particles for which spin-parity (3/2 +) has been established appear to form a decuplet pattern. The charge multiplets required have already been set out in terms of quarks in *Figure 8-5*. If these states are re-plotted in terms of the hypercharge Y and isospin component I_3, they form the array shown in *Figure 9-5* which displays the three-fold symmetry expected with SU (3) symmetry. The mean masses for each charge multiplet are 1236 *MeV* for Δ, 1382 *MeV* for Σ^* and 1530 *MeV* for Ξ^*. These masses increase approximately in a linear progression with increasingly negative Y, in steps of about 147 *MeV,* and this linear dependence on Y is just what the mass formula (8.6) would require for these charge multiplets (in terms of this formula, the Σ^* has mass ($M_0 + 2b$) and the mass interval between the states is $- (a + 3b/2)$). Irrespective of the validity of this formula, this pattern of states clearly required for its completion the existence of a tenth (3/2 +) particle, with mass about 1680 *MeV,* a charge singlet particle with strangeness s = $-$ 3 and with charge $Q = -$ 1, as was first pointed out by Gell-Mann in 1962. Although there still has not been a spin-parity determination for the Ω^- particle, its discovery in an experiment at Brookhaven in 1964, with all of the other quantum numbers correct and with mass (1672 MeV) so close to that required, carried instant conviction that this was indeed the missing decuplet particle and led to a widespread acceptance of the physical ideas underlying the $SU(3)$ symmetry.

There is now beginning the study of a new spectroscopy for these subnuclear energy levels. There are many transitions observed to occur between them, quite analogous to the nuclear transitions which give rise to nuclear spectroscopy, and the systematics of these transitions will have to be examined and understood. These transitions may involve π, K or η emission through the strong interactions: several strong π transitions which have been reported are as follows,

$$\Lambda^*(1815) \rightarrow \Sigma^*(1385) + \pi, \qquad (9.12)$$
$$\Sigma^*(1765) \rightarrow \Lambda^*(1520) + \pi, \qquad (9.13)$$
$$\Sigma^*(1660) \rightarrow \Lambda^*(1405) + \pi, \qquad (9.14)$$

The γ-transition

$$\Lambda^*(1520) \rightarrow \Lambda + \gamma \qquad (9.15)$$

has also been measured recently. Weak decay transitions will not be competitive for any of these particles other than the $(\frac{1}{2} +)$ baryon octet and the Ω^- particle. However, the excitation of some of these particles by high-energy neutrinos is a possibility (limited to nucleon targets, unfortunately); the transition

$$\nu_\mu + P \rightarrow \Delta^{++} + \mu^- \qquad (9.16)$$

has already been reported, and we may hope to see some strangeness-changing weak transitions leading to Σ^* particles, such as

$$\bar{\nu}_\mu + N \rightarrow \Sigma^{*-}(1765) + \mu^+, \qquad (9.17)$$

when new neutrino experiments become possible with the large hydrogen and deuterium bubble chambers whose construction is now nearing completion at the Brookhaven, CERN, and Argonne laboratories.

As our knowledge of these particles has been increasing, it has appeared possible that the observed $SU(3)$ multiplets may themselves be grouped into still larger supermultiplets which may be said to reflect a *"charge, hypercharge and quark-spin independence"* of the underlying superstrong forces. The essential difference between the decuplet states set out in *Figure 8-5* and the baryon octet states (not shown here in terms of quark states) is that the three quark spins must add to total spin $\frac{1}{2}$ for the octet, whereas they add to total spin 3/2 for the decuplet. With this additional quark-spin independence, the $(\frac{1}{2}+)$ baryon octet and the $(3/2+)$ decuplet states come together within one larger multiplet, having the same internal structure with respect to all other variables; all of these 56 states (counting the $(2S+1)$ spin states for each particle we have $56 = 2 \times 8 + 4 \times 10$) would then have the same interactions and the same energy with respect to the superstrong forces. This larger symmetry is generally referred to as $SU(6)$ symmetry ($6 = 3 \times 2$, where 2 is the number of states for quark spin $\frac{1}{2}$). A typical measure of the $SU(6)$-breaking forces is given by the mass difference between this octet and decuplet, which is at most 300 MeV, very little larger than the measure we have given previously for the $SU(3)$-breaking forces.

Next, it appears quite possible that the internal structure of these baryonic particles may also involve some internal orbital angular momentum L, which must be coupled with the total quark spin S to give the total spin J for the particle. Many of these excited states might then simply represent rotational excitations of the particle states lying lower in mass. There could also be internal radial motions of excitation, which would not contribute angular momentum to the particle spin J. With these possibilities, the $SU(3)$ unitary multiplets may be grouped together in large *supermultiplets* characterized by a definite $SU(6)$ symmetry α, an internal angular momentum L with parity w, and an index n specifying the degree of radial excitation.

For clarity, the nucleonic members of the $SU(3)$ multiplets shown on *Figure 9-4* have been plotted separately on *Figure 9-6*, the particles with negative parity being separated from those with positive parity. At present, all of the nucleonic particles below mass 2 GeV can be accommodated within four of these (α, Lwn) supermultiplets. The significance of these large supermultiplets is far from being understood at present. Much work is still to be done to check whether the details of the particle spectrum can be understood in terms of a small number of $SU(6)$-breaking interactions and whether all of the transitions experimentally

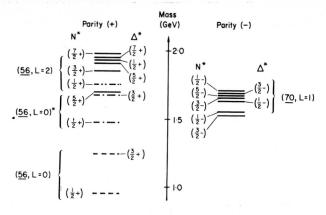

*Figure 9-6. All of the nucleonic states are plotted separately to indicate
the bands in which they appear. All negative parity baryons are plotted
on the right, all positive parity baryons are plotted on the left, and the
states are separated according as their isospin is 1/2 (N* states) or 3/2
(Δ* states). The negative parity particles below 2 GeV appear to belong
to a SU(6) 70 super-multiplet with orbital angular momentum L = 1 and
negative parity. The positive parity states appear to belong to three SU(6)
supermultiplets, the upper two of which can be interpreted as a radial
and a rotational excitation of the ground state supermultiplet.*

observed do fit the patterns expected for them when they occur
either within or between these supermultiplets.

The idea that the high-lying particles may arise as rotational
excitations of the particles with lowest mass has been put forward
by Chew quite independently of these supermultiplet interpretations.
On rather general grounds, based on some ideas about states of
higher angular momentum put forward earlier by Regge, Chew
argued that a low-lying particle with spin S should give rise to a
series of "Regge recurrence" particles with spins $J = (S + 2)$,
$(S + 4)$, $(S + 6)$, . . . increasing in steps of two with mass
values lying on a smooth curve as function of J. This can be
conveniently tested for one case in particular; $\pi^+ P$ interactions can
excite only Δ particles, and very accurate total cross-section
measurements have led to the identification of a long series of
prominent Δ particles. Spin values have been determined only for
the $\Delta(1236)$, $\Delta(1950)$ and $\Delta(2420)$ particles, and these have
plotted in *Figure 9-7* as a graph of spin J vs. (mass)^2 (known as a
Chew-Frautschi plot). The first three points lie fairly well on a
straight line and it is remarkable that the two further Δ particles

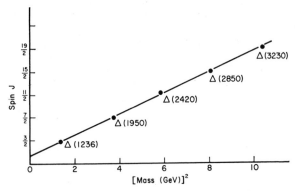

Figure 9-7. Spin J is plotted against (mass)² (a Chew-Frautschi plot) for the prominent Δ resonant states. The spin values of the upper two states have not yet been determined experimentally, but if they are plotted so as to fall on the straight line, then they have the spin values which follow next after the sequence 3/2, 7/2, and 11/2.

which are known, Δ(2850) and Δ(3230), also lie on the extension of this straight line, provided we assume that their spin values continue the uniform progression 3/2 +, 7/2 +, 11/2 +, . . . in steps of 2 units. This requires spin-parity values (15/2 +) for Δ(2850) and (19/2 +) for Δ(3230), uncommonly large values, and it will be very interesting to see direct measurements for these spins in due course.

A parallel development has been going on for higher mesonic particles. These are all unstable particles, and their analysis has therefore been relatively difficult. The particles best known are those named the "vector mesons", whose spin-parity value is (1 −). They are shown on *Figure 9-8* which plots all the mesons established up to mass about 1600 *MeV*, the $\rho(765)$ charge triplet, the $K^*(890)$ and $\overline{K}^*(890)$ charge doublets, and the $\omega(783)$ and $\phi(1019)$ charge singlets. They are set out on the (I_3, Y) plot on *Figure 9-9*. There are nine states altogether which we would naturally expect to consist of an octet and an $SU(3)$ singlet. We can use the mass-formula for vector mesons, analogous to Eq. (8.8), to calculate the mass of the $I_3 = Y = 0$ meson V_0 in the octet,

$$3\ m^2(V_0)\ =\ 4\ m^2(K^*)\ -\ m^2(\rho). \qquad (9.18)$$

This gives a mass of 930 *MeV*, which agrees with neither $\omega(783)$

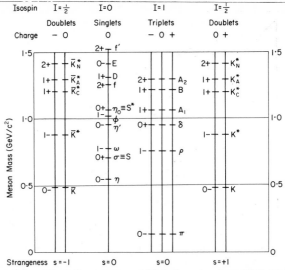

Figure 9-8. All of the mesons established up to about 1500 MeV are plotted in columns according to their charge, isospin, and strangeness. The states which are (0 −) all belong to an octet and a singlet. The states (1 −) form a nonet, as do also the states (2 +).

nor $\phi(1019)$ but sits right in between them. The answer turns out to be that the octet state V_0 and the unitary singlet $V(1)$ become mixed together by the strong interactions which break $SU(3)$ symmetry. The only sensible thing to do is to refer to these nine vector states as forming a nonet. We have mentioned

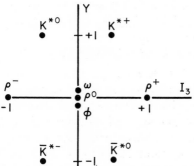

Figure 9-9. The nine vector meson states are plotted on a (I_3, Y) plane. They form a mixed (octet plus singlet) state with respect to SU(3) symmetry. These states are usually described as a nonet.

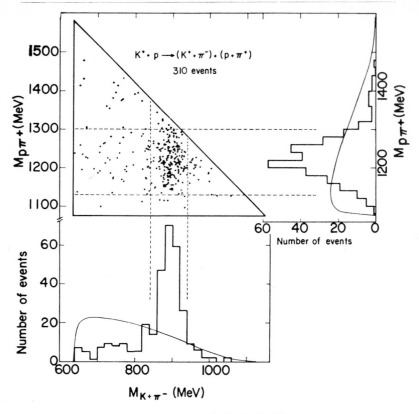

Figure 9-10. Each event of the type (9.19) for K⁺ laboratory momentum 1960 MeV/c is represented by one point on this plot of M(Pπ⁺) vs. M(K⁺π⁻). By projecting on each axis, plots of the mass values are given individually. The plot shows that K*(890) production is generally accompanied by Δ⁺⁺(1236) production. The triangle indicates the kinematically allowed region.

already that the quark-antiquark picture leads naturally to nine meson states; the nine vector mesons correspond with the nine states shown in *Figure 8-4* after the antiquark spin has been turned to be parallel with the quark spin in each case. *Charge-, hypercharge- and quark-spin independence* (i.e., *SU*(6) symmetry) for the superstrong forces would give the same mass to all these nine vector mesons and also to the (0 −) octet mesons. The η′ meson is then an *SU*(6) singlet state, with zero charge, zero hypercharge and zero spin; it could then be quite distinct in mass

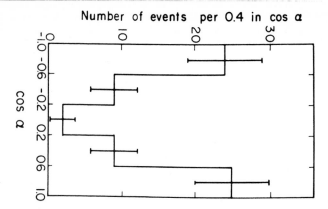

Figure 9-11. The angular distribution for the decay $K^*(890) \to K^+\pi^-$, as seen in this K^* rest frame, relative to the direction of the incoming K^+ meson, is plotted for the K^+P reaction (9.19) at 1960 MeV/c in terms of $\cos\alpha$ and fits well a $\cos^2\alpha$ form.

from the other $(1 \,—)$ and $0 \,—)$ mesons just mentioned, although its mass does still lie quite close to the other mass values.

To illustrate the methods used for the spin-parity determination of the unstable mesons, let us consider the reaction

$$K^+ + P \to (K^+ + \pi^-) + (\pi^+ + P) \qquad (9.19)$$

induced by 1960 MeV/c K^+ mesons from the Berkeley Bevatron. For each event, the energies $M(K^+\pi^-)$ and $M(\pi^+P)$ are calculated from the measured momenta and energies, using the form equivalent to Eq. (9.9),

$$M(ab)^2c^4 = (E_a + E_b)^2 - c^2(\mathbf{p}_a + \mathbf{p}_b)^2, \qquad (9.20)$$

and the event is represented by a dot on *Figure 9-10*, with coordinates $(M(K^+\pi^-),\ M(\pi^+P)$. We note that these dots cluster strongly in the region where $M(K^+\pi^-)$ has value about 900 MeV and $M(\pi^+P)$ has value about 1220 MeV. The interpretation of this scatter plot is that the majority of the events occur through K^* and Δ excitation, followed by their decay to give the final state (9.19),

$$K^+ + P \to K^{*0} \qquad\qquad + \triangle^{++}$$
$$ \big\downarrow \qquad\qquad\qquad \big\downarrow$$
$$ \to K^+ + \pi^- \qquad\quad \to \pi^+ + P \qquad (9.21)$$

The angular distribution of the decays is then examined. We consider the K^* decay and determine the angle α between the final K^+ momentum and the ingoing K^+ momentum as seen by an observer sitting on the K^{*0} particle. The distribution found for *cos* α *is* shown in *Figure 9-11.* This is quite distinctive and is well fitted by the form $cos^2\alpha$, from which it is possible to conclude that there is just one unit of orbital angular momentum in the final state from $K^{*0} \rightarrow K^+\pi^-$. Since both K^+ and π^- mesons are spinless, angular momentum conservation requires that the K^{*0} meson has spin 1; similarly, since K^+ and π^- each have parity $(-)$, the K^{*0} parity is simply that for the final orbital motion[†], which gives the result $(-)$.

The particular importance of the ρ^0, ω and ϕ mesons is that they have the same quantum numbers Q, B, Y and spin-parity as does the photon. Hence they play an important role in all photon-induced processes, for example, in the theoretical interpretation of the electric and magnetic structure of the proton and neutron. They can be produced relatively easily in photon interactions. One particularly interesting process is a kind of diffraction process which occurs for photons incident on a heavy nucleus, without disturbing the nucleus, for example

$$\gamma + Pb^{208} \rightarrow Pb^{208} + \rho^0. \qquad (9.22)$$

From the analysis of this process, one can derive the total cross-section for a ρ-meson incident on a nucleon, a quantity whose determination nobody would have considered even remotely feasible several years ago.

Another production process of special interest is vector meson production in electron-positron collisions. For these studies, a special kind of accelerator has been devised and constructed, the intersecting storage ring or colliding beams accelerators which will be described in Professor Panofsky's lectures. Since the photon coupling is small (measured by the fundamental dimensionless constant $e^2/\hbar c = 1/137$), the most probable processes are those which involve only one intermediate photon. The vector meson production processes are

$$e^+ + e^- \rightarrow \gamma \rightarrow \rho^0, \omega \text{ or } \phi. \qquad (9.23)$$

The interesting feature is that this process forms these vector mesons at rest in the laboratory, so that the experimental study of their decay processes can now be carried out very much more conveniently and accurately.

Another nonet appears among the states on *Figure 9-8,* con-

---

† The general rule is that for simple orbital motion with L units of angular momentum, the associated parity is given by $(-)^L$.

sisting of the following particles with spin-parity $(2+)$, the charge triplet $A_2(1300)$, the strange doublets K_N^* (1400) and $\overline{K}_N^*$ (1400), and the charge singlets $f(1260)$ and $f'(1515)$. For the other spin values with parity $(+)$, the complete octets or nonets are not yet known. However, it appears rather likely that all of these $(+)$ parity mesons will turn out to consist of a series of nonets which form together another *supermultiplet* characterized by the same $SU(6)$ symmetry as the low-lying $(-)$ parity mesons, but with an internal orbital angular momentum $L = 1$ with $(-)$ parity. At least none of the mesons known in this mass region are in conflict with this interpretation. But there are still quite a number of mesons still to be found if this picture is true and there could be a lot of surprises ahead for us!

Many further mesons are known for higher mass values, as high as 2400 *MeV*. One particularly illuminating experiment has been carried out at CERN with an instrument called the "missing-mass spectrometer". This is essentially an analysing system which can measure the protons emerging from a hydrogen target exposed to a negative meson beam, with high precision and high statistics. From the knowledge of the incident beam momentum and the final proton momentum, we can deduce the missing energy E and the missing momentum P, and hence the "missing mass" $\sqrt{(E^2 - c^2P^2)}$, for the reaction

$$\pi^- + P \rightarrow P + (\text{Missing Mass})^- \qquad (9.24)$$

The (Missing Mass)$^-$ represents some mesonic state, with hyper-charge zero and at least charge triplet, irrespective of its mode of decay. If the mesonic particles were all long-lived and therefore well-defined in mass, the distribution of missing masses would just reflect the mass spectrum of these particles. However, most of them are very unstable, increasingly so with increasing mass value, and these broad levels will overlap strongly, producing a smooth background "missing mass" distribution. The spectro-meter could only hope to be sensitive to the relatively long-lived particles, and it seemed unlikely that any particles with such long lifetimes should exist with high mass values.

The mass spectrum shown in *Figure 9-12* was determined at CERN in 1966. It shows most striking structure in detail, and this may well be described as the Franck-Hertz experiment for

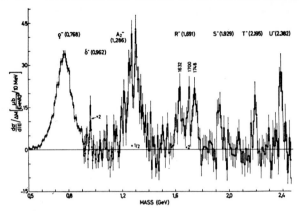

Figure 9-12. The mass spectrum for non-strange mesons with charge Q = — 1, as determined with the "Missing Mass Spectrometer" at CERN, for the reaction (4.24).

the subnuclear world. The particles already well-known, the ρ and A_2 particles, appear strongly as broad peaks, and there are many narrow peaks appearing right up to the highest mass region explored. The power of the instrument is well illustrated by the narrow peak δ at 962 *MeV*; it was only at the end of 1968 that the existence of this particle has become confirmed by other experimental techniques. The high-mass peaks have not yet been confirmed in detail by other techniques, but the experimental work which has been stimulated by this data (and which usually investigates only particular modes of decay for the mesons produced) has confirmed that there do exist quite long-lived unstable mesons at these high mass values. For example the *S, T* and *U* mesons shown on *Figure 9-12* all lie above the threshold for the proton-antiproton system, and there is the possibility of exciting them directly using antiproton beams incident on a hydrogen target,

$$\overline{P} + P \rightarrow (\text{Meson})^{*0} \rightarrow \overline{P} + P, \qquad (9.25a)$$
$$\searrow \text{lighter mesons.} \qquad (9.25b)$$

Some of these experiments have been carried out recently (1969), and they appear to confirm the existence of these high-mass mesons although these observations indicate rather larger values for their lifetime widths than does *Figure 9-12*. In particular, experimental observations on the cross-sections for $\overline{P}$ incident on N (in deuterium) show a strong indication of the S^- meson shown on

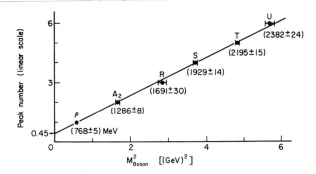

Figure 9-13. The order of the peaks shown in Figure 9-12 is plotted against (mass)² for each dominant peak or group of close peaks. It is shown that these points lie on a straight line, and the plot is interpreted as a Chew-Frautschi plot in which the ordinate gives the spin of the corresponding particle. This straight line crosses the vertical axis at 0.45, and joins quite smoothly with the continuation of this "Regge-trajectory" to negative values of (mass)².

Figure 9-12, but this work has not progressed far enough yet to give any spin-parity values for these mesons.

The relatively long lifetimes (or small lifetime widths Γ) observed for these high-mass mesons came as a complete surprise to us all (although the lifetimes are still of order 10^{-23} sec). The only way we have been able to understand them so far is by supposing that the mesons have increasingly high spin values as we go up the sequence, ρ, A_2, R, S, T, U . . . A particle of high spin has to get rid of a large amount of angular momentum in making transitions to the lower-mass particles which have lower spin values, and the difficulty for it to do this may be sufficient to account for its relatively high stability. This interpretation fits in with the observation made by the CERN workers that the (mass)² values for these mesons lie approximately on a straight line, as shown on *Figure 9-13*, when plotted against their order in the sequence of prominent resonances on *Figure 9-12*. Since spin (1 −) is known for the ρ particle, and (2 +) for the A_2 particle, the natural interpretation is that *Figure 9-13* is actually a Chew-Frautschi plot so that the ordinate gives the spin value for the corresponding meson. This picture requires spin-parity (3 −) for R, (4 +) for S, (5 −) for T, (6 +) for U, and so on. This sequence of mesons would then represent a series of rotational excitations of the ρ meson, and this line on *Figure 9-13* would

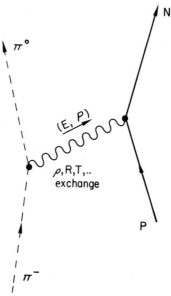

Figure 9-14. A graphical representation of the exchange of a virtual meson carrying energy E and momentum P from the incident pion to the target nucleon. Only mesons with I = 1 (or more) and odd spin (and capable of decaying to two pions) can be exchanged in this charge-exchange process.

then be referred to as the "ρ Regge-trajectory" after the theoretical work of Regge which first introduced the idea that particles of different spin might be related and their (mass)^2 values connected by a simple smooth curve (although the idea of a series of rotational excitations based on a particular ground state is a very old one in molecular physics, of course).

What is particularly interesting about the Regge-trajectory in elementary particle physics is that it also has physical significance to the left of the origin, for negative values of (mass)^2. This is the case for those scattering and reaction processes which are believed to be mediated at very high energies dominantly by the exchange of $I = 1$, $Y = 0$ mesons, such as the charge-exchange process (see *Figure 9-14*).

$$\pi^- + P \rightarrow \pi^0 + N. \tag{9.26}$$

In the reference frame in which the two initial (or final) particles of reaction (9.26) have total momentum zero, the pion and neutron go out with the same momentum values as the ingoing pion and proton had, but with a change of direction. Hence there is a momentum transfer P from the pion to the nucleon but no energy change ($E = 0$); the (mass)$^2c^4 = (E^2 - c^2P^2)$ which we associate with the meson exchanged between the pion and nucleon has a negative value, reaching the value zero for pion scattering angle zero. The form of the scattering angular distribution for reaction (9.26) is then controlled by the extension of the curve on *Figure 9-13* to negative (mass)^2 values, and the energy-dependence of the cross-section for scattering at $\Theta = 0^\circ$ is controlled by the value of J (the ordinate on *Figure 9-13*) at (mass)$^2 = 0$. It is satisfactory to find that the analysis of the charge-exchange scattering data leads to a "Regge-trajectory" for (mass)$^2 \leqslant 0$ which does join quite smoothly with the line for (mass)$^2 \geqslant 0$ shown on *Figure 9-13*, the value $J = 0 \cdot 5$ being given at (mass)$^2 = 0$ by the scattering analyses.

We must admit here that there is rather little evidence yet to support this story about the *R, S, T, U,* . . . mesons, except for some strong indications that one meson with mass about 1700 *MeV* does have spin-parity (3 −) (but it is not necessarily the *R*-meson). There are many experiments of various kinds now under way to learn more detail about the properties of these high mass mesons. This will involve a spectroscopy for the mesons, with a great complexity of transitions between them, whose patterns must be understood. A few examples of transitions already observed are

$$\eta' \ (950) \quad \rightarrow \rho + \gamma, \quad\quad\quad\quad (9.27a)$$
$$\text{``}R(1640)\text{''} \rightarrow f + \pi, \quad\quad\quad\quad (9.27b)$$
$$\text{``}R(1700)\text{''} \rightarrow \rho + \rho \text{ and } \omega + \pi, \quad\quad (9.27c)$$

where the notation *"R"* refers generically to a meson whose mass value lies in the *R* band of *Figure 9-12* — the two mesons R(1640) and R(1700) are definitely distinct particles. Meson spectroscopy is still in a relatively primitive state, despite a very great deal of hard work by the experimenters. The mesons can be studied only in processes where they are both produced and decay. They can be recognised only by accurate measurements on their decay products, and most of the mesons have a large number of alter-native decay modes. Many of the results obtained by different bubble chamber groups about the same meson (as seen in different production processes, generally) are contradictory, and our picture about these higher mesons is very far from clear; almost the only thing we are really sure about is that such mesons do exist. Meson spectroscopy is bound to become a more highly developed field of work in the future, probably involving the use of new experimental techniques.

Now we may look back at *Figure 2-1*. We see that we have now reached a situation very similar to that we have already known in the study of atoms and nuclei. The main difference lies in the energy scale involved, the unit is about 1 electron volt for the atomic excitations, about one million electron volts for the nuclear excitations, and now about one hundred million electron volts for the subnuclear excitations. Just as the energy level structure for atoms and for nuclei suggested that these objects were composite and had internal structure, the complex energy level patterns observed for the hadrons also seem to indicate that there must be substructure in the hadrons, and that there must exist superstrong forces operative within this structure which have some very high degree of symmetry not yet understood.

CHAPTER 10
(R. H. Dalitz)

The Weak Decay Interactions

The interactions responsible for the decay processes observed for the semi-stable hadrons, the lightest of the mesons and the baryons, are exceedingly weak relative to the strong, nuclear forces we have just been discussing. This is well illustrated by comparing the lifetimes for these decay processes, typically 10^{-8} sec to 10^{-10} sec, with the characteristic nuclear time of 10^{-23} sec. The strengths of these interactions is measured conventionally by a coupling parameter named G, whose value is

$$G = (1 \cdot 0272 \pm 0.0002) \times 10^{-5}/M_P{}^2, \qquad (10.1)$$

where M_P denotes the proton mass.

The most surprising aspect of the weak interactions has been the discovery that they violate a number of conservation laws and symmetry principles which had always been thought quite secure since they were concerned with symmetry properties of space-time. We have already mentioned the τ-θ dilemma, where suspicion of these long-sacred principles first arose, the fact that there existed two competing decay modes (2π and 3π) which appeared to come from the same K^+ meson but whose final states definitely had opposite parity. However, no matter how difficult it was to avoid this conclusion, these phenomena did not demonstrate the failure of parity conservation in an explicit way. It was Lee and Yang who enlarged this suspicion to the weak decay interactions in general, by asking what evidence there was that any of these interactions were invariant under the parity operation P (space reflection). When it became clear that there was no such evidence, the famous experiment on the beta decay of polarized Co^{60} was carried out by Miss Wu at the National Bureau of Standards in Washington to provide an explicit test for P-

invariance. We have already illustrated this experiment in *Figure 7-1*, and we have discussed from first principles in Chapter 7 how *P*-invariance directly requires that the intensity of emission of beta particles along the axis of spin polarization for the Co^{60} should be the same, both parallel and antiparallel to the direction of this spin. In fact, the measured intensities along and opposite the spin direction for the polarized Co^{60} were not equal, and the failure of parity conservation for the beta-decay interaction was made quite explicit.

The failure of particle-antiparticle conjugation *C* for beta-decay was soon demonstrated in an equally unambiguous way for the beta-process giving rise to muon decay. This is illustrated in *Figure 10-1*. Owing to the failure of *P*-invariance in the beta-decay interactions, the positrons emitted in the decay of unpolarized μ^+ particles have a definite spin about their direction of emission (i.e., a longitudinal spin). This longitudinal polarization can be measured in a number of ways which depend only on the validity of *P*-invariance for electromagnetic interactions (which has already been well-tested for a long time). We shall just indicate briefly one method which measures the polarization sign readily and which can be used for e^- as well as for e^+. When the electrons or positrons strike a target, they emit photons because of their collision with it and some of their longitudinal polarization is passed on to these photons as circular polarization. This circular polarization is determined by measuring how many of these photons can pass through a block of magnetised iron (which contains electrons whose spin directions are the same as the direction of magnetisation) when the magnetisation direction is parallel, and when it is opposite, to this beam direction; the existence of an effect depends on the fact that the scattering of circularly polarized photons by the polarized electrons in the iron depends on the relative direction of these two polarizations in a known way. This procedure has led to the conclusion (consistent with the results from other experimental procedures) that unpolarized μ^+ decay to give positrons with a right-handed spin along their direction of emission (as shown in *Figure 7-1*), whereas unpolarized μ^- decay to give electrons with a left-handed spin.

This empirical result contradicts *C*-invariance directly. The left

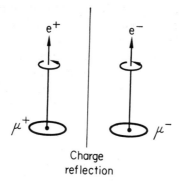

Charge
reflection

*Figure 10-1.   The left figure depicts the e^+ emitted in the decay of an
unpolarized μ^+ meson, showing that it spins with a right-handed helicity
relative to its direction of emission.   The operation of particle-antiparticle
conjugation (i.e., charge reflection C) transforms this situation into the
picture on the right.   Invariance with respect to the operation C would
require that the e^- should also spin right-handed about its direction of
momentum, when emitted from an unpolarized μ^- meson.   This is
contrary to the observation that the e^- thus emitted actually spin
left-handed.*

half of *Figure 10-1* illustrates μ^+ decay, in accordance with the
empirical result.   The right half shows the situation after carrying
out the operation of particle-antiparticle conjugation C.   This
operation does not affect the spin, nor the direction of motion of
particles, but simply replaces each particle by its corresponding
antiparticle, so that the right half of *Figure 10-1* depicts our
expectation for the spin polarization for the electrons emitted in
the decay of unpolarized μ^- particles, if C-invariance were to hold
good for the muon beta-decay interaction.   In fact, the experimental result is exactly the opposite of this expectation; the e^-
from unpolarized μ^- decay spin in the opposite sense to that for the
e^+ from unpolarized μ^+ decay.

The observation of parity nonconservation for the beta-
interaction was in direct conflict with our intuitive ideas about
space in that it appeared to mean that our physical laws would
take a different form in the mirror world (where the reference
axes were left-handed) from their form in our world (where the
reference axes are right-handed, by convention).   This situation
would enable us to make an absolute distinction between the two

worlds, although intuitively we have great difficulty in understanding why Nature should prefer one of these worlds to the other, in an absolute sense. However, Landau pointed out that it was quite possible that C- and P-invariance could be violated in such a way that the laws of physics were still invariant under the joint operation of C and P, i.e., CP-invariant. In this situation, the laws of physics would have the same form for particles in our world as for the corresponding antiparticles in the mirror world. Unless we had some *absolute* means to decide whether we were dealing with particles or with antiparticles, and we do not within the laws of physics at this stage in these lectures, then we still could not use these phenomena of P-violation to distinguish absolutely between our world and the mirror world.

Since CPT-invariance, i.e., invariance under the combined operations of C, P and T, is believed to hold valid under rather general conditions, Landau's proposal for CP-invariance would hold rather naturally if the invariance T with respect to time-reversal held for all interactions. Then, with CPT- and T-invariance, the invariance with respect to

$$(CPT)T = CP(T^2) = CP \qquad (10.2)$$

would necessarily hold, since two successive applications of the operation of time-reversal bring us back to the initial situation.

We shall now consider more detailed questions about the nature of the weak decay interactions. Let us consider first those decay processes which involve leptons. The pattern of these interactions is shown on *Figure 10-2*. The simplest process is the muon decay which corresponds to the horizontal link between (μ^+, ν_μ) and (e^+, ν_e),

$$\mu^+ \rightarrow \bar{\nu}_\mu + (e^+ + \nu_e), \qquad (10.3)$$

and the interaction takes the form of the product of two currents, one associated with the (μ, ν) system, the other with the (e, ν) system. The amplitude for this μ^+ decay process is the fundamental constant G given in (10.1).

The link on the right generates all of the beta-decay transitions known. The hadron current can connect hadrons with the same strangeness ($\Delta s = 0$) or with strangeness different by one unit ($\Delta s = \pm 1$). Examples of these processes are the neutron beta decay

$$N \rightarrow P + e^- + \bar{v}_e \qquad (10.4)$$

for $\Delta s = 0$, and the Λ beta-decay

$$\Lambda \rightarrow P + e^- + \bar{v}_e \qquad (10.5)$$

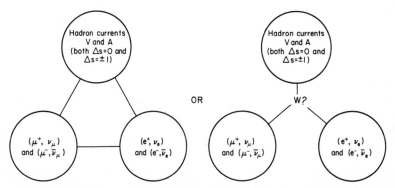

Figure 10-2. Three classes of weak current are known. Their combinations form three distinct weak decay processes involving lepton pairs. A simple interpretation of this degree of order among the weak interactions is that the weak interaction may be carried by a universal intermediate vector meson W which couples with equal strength to each of these currents.

for $\Delta s = + 1$. It has been found generally that the decay interactions involving $\Delta s = \pm .1$ are weaker than those with $\Delta s = 0$ by a factor of about 20, after due allowance has been made for the effect of the magnitude of the energy released in the decay process. The free neutron decay (10.4) is slow simply because the energy released in the $N \rightarrow P$ transition is only $1 \cdot 3$ *MeV*; if the $\Lambda \rightarrow P$ transition (energy release 177 *MeV*) had the same amplitude as that for the $N \rightarrow P$ transition, the decay rate predicted for the process (10.5) would be about 20 times greater than is actually observed.

Although the $\Delta s = 0$ beta-decay process for the Σ^+ particle

$$\Sigma^+ \rightarrow \Lambda + e^+ + v_e \qquad (10.6)$$

is well-known, the $\Delta s = + 1$ beta-decay process

$$\Sigma^+ \rightarrow N + e^+ + v_e \qquad (10.7)$$

has not been seen. This is an example of a general rule, called the $\Delta s / \Delta Q = + 1$ rule. This states that a hadronic beta-decay process will not occur unless the total charge change for the hadrons involved ($\Delta Q = -1$ for process (10.7)) has the same sign as their total strangeness change ($\Delta s = +1$ for (10.7)).

These decay processes can also be described as due to the interaction of a hadronic weak current with the (e, ν_e) current. Since parity is not conserved in these beta-decay processes, the hadronic current naturally has two parts with opposite parities, which we shall denote by $V(\Delta s)$ and $A(\Delta s)$, since they depend on the strangeness change Δs. The symbol V stands for "vector" and the $V(0)$ current is, in fact, closely connected with a part of the electromagnetic current for the hadrons; the symbol A stands for "axial vector", which just means a spatially-directed quantity whose parity is $(+)$†. The $\Delta s = -1$ currents are not independent of the $\Delta s = +1$ currents, but are completely determined when the latter are known in detailed form. A remarkable synthesis of all these hadronic currents was achieved by Cabibbo a few years ago when he proposed that these various currents were related to each other in just the same way as would be the corresponding members of a unitary octet of mesons, but with a strength which depended on the Δs involved. He wrote the hadron currents in terms of one free parameter, the "Cabibbo angle" θ, as follows:

$$J_{wk}(\triangle Q = +1) = \{\cos\theta\,(V(0) + A(0)) + \sin\theta\,(V(+1) + A(+1))\}$$
(10.8)

where its coupling amplitude to the (e, ν_e) current is denoted by G_h. The relative weakness of the $\Delta s = +1$ transition corresponds to a relatively small value for θ, given consistently as about $0\cdot24$ radians from the analysis of a number of different beta-decay phenomena. Most remarkably, the amplitude G_h is then found to be equal to the amplitude G for muon decay, within the accuracy of the data. At an earlier stage in this subject it had been conjectured as a hypothesis of universality for the weak interactions that the amplitude G_V for $V(0)$ might have the same value as G,

---

† To give an illustration from classical mechanics, the position vector **r** and the momentum vector **p** are examples of V vectors, since they reverse when the axes are reflected in the origin. On the other hand, the angular momentum vector **r** × **p** is a spatially-directed quantity which remains unchanged when the axes are reflected and **r**, **p** become replaced by − **r** and − **p**, and so provides an example of an A vector.

but as the experimental data became more accurate and more certain, the gap between G_V and G became too large (relative to the experimental uncertainties) to be ignored. With Cabibbo's suggestion, we see that actually $G_V = G_h \cos \theta$, where the hypothesis of universality is now given by $G_h = G$; G_V is therefore predicted to be about 3% smaller than G, and this is in excellent agreement with the facts.

The remaining link in the left figure of *Figure 10-2* connects the (μ, ν_μ) current with the hadron current (10.8). This gives rise to the $\Delta s = 0$ capture process

$$\mu^- + P \to N + \nu_\mu \qquad (10.9)$$

and $\Delta s = + 1$ decay processes such as

$$\Sigma^- \to N + \mu^- + \bar{\nu}_\mu. \qquad (10.10)$$

The hadron current also induces the $\Delta s = 0$ transition "$\pi^- \to$ nothing" and the $\Delta s = + 1$ transition "$K^- \to$ nothing", which give rise to their dominant decay processes.

$$\pi^- \to \mu^- + \bar{\nu}_\mu, \qquad (10.11a)$$

$$K^- \to \mu^- + \bar{\nu}_\mu. \qquad (10.11b)$$

Our knowledge of all these muonic processes can be summed up quickly; the amplitudes for a muonic process and the corresponding electronic process are identical, except for the explicit change of the lepton mass wherever it occurs. For example, the amplitudes (10.11) have the lepton mass m_μ as a factor (this actually follows from the V, A) form for the weak currents): the corresponding processes $\pi^- \to e^-\bar{\nu}_e$ and $K^- \to e^-\bar{\nu}_e$ have amplitudes with the factor m_e, so that they occur at a much lower rate than the muonic processes (10.11).

Here we should interpose a few remarks about the lepton currents (e, ν_e) and (μ, ν_μ), which also have parts with both parity signs. All of the experiments indicates the form of these currents to be just such that the neutrino with right-handed spin along its direction of motion (the state of positive helicity) and the antineutrino with left-handed spin along its direction of motion (the state of negative helicity) do not participate in the weak interactions. This may be stated more firmly: *the neutrino has only a negative helicity state, and the antineutrino has only a positive helicity state.* As far as physics goes, the states of opposite helicities to these do not exist, since they do not participate in any physical processes. The v and $\bar{v}$ states which do interact have opposite helicities, so that these states are not C-conjugate; hence these interactions must violate C-invariance. In the mirror world, the space-reflected neutrino has positive helicity, yet does interact; hence the weak-interactions must automatically violate P-invariance. However, we must emphasize that not all of the P- and C-violating effects are connected with the properties of the neutrino although these do illustrate the violation of C and P rather nicely.

Now, with the hadronic current of Cabibbo, we see that the three links of *Figure 10-3* each have the same form and the same strength; in this sense *the weak interaction is universal and has the (V, A) form.* This would be readily understood if there existed some intermediate field W which carried the weak inter-action between the various currents. The existence of (V, A) couplings would require W to be a vector particle (spin 1 $-$). It must have charge $+ e$ (with an antiparticle $\overline{W}$ of charge $- e$). Its coupling to the three types of weak interaction current must be universal (at least for the V terms) with strength f (analogous to the electric charge e) given by

$$G = (f^2/\hbar c)/M_W{}^2 \qquad (10.12)$$

This hypothesis that there should be an *intermediate vector boson* with a universal weak interaction is very attractive, since it auto-matically organises the weak interactions into the pattern we have observed so far. If we were to suppose further that W was coupled primiraly to a quark triplet, then the only strangeness-changing interaction would be $p \to \lambda + W^+$, which involves $\Delta s = -1$ and $\Delta Q = - 1$, thus permitting only $\Delta s/\Delta Q = + 1$ tran-sitions among the composite physical particles. However, it is not necessary to accept such a drastic hypothesis in order to account for the $\Delta s/\Delta Q$ rule; this rule also follows automatically from Cabibbo's hypothesis (10.8), since the only elements of an octet which have both strangeness and charge are the K^+ and K^- states and (charge/strangeness) has the value $+ 1$ for both of these.

No direct evidence has yet been found for the existence of this W boson despite considerable search. All that is known is that its mass cannot be less than 2 GeV (see below). Its decay rate will be proportional to f^2, rather than to the G^2 characteristic of other weak decay processes, so that its lifetime is likely to be roughly a geometric mean between the lifetimes typical of the usual weak decay processes mediated by W ($\sim 10^{-10}$ sec) and the characteristic nuclear time ($\sim 10^{-23}$ sec), thus about 10^{-17} sec. Of course, we shall not know the coupling constant ($f^2/\hbar c$) until the mass M_W is determined. It is interesting to remark that the mass cannot be arbitrarily large: according to (10.1) and (10.12), the coupling constant $f^2/\hbar c$ would reach a value of order $1/137$ already for mass M_W about 20 GeV.

In recent years, direct evidence on the neutrino and its properties has been obtained from the study of reaction processes initiated by neutrinos and antineutrinos, measurements which most of us had not expected to see within our lifetime. We shall discuss these briefly:

(i) Cowan and Reines established the existence of $\bar{\nu}_e$ by observing the reaction

$$\bar{\nu}_e + P \rightarrow N + e^+ \qquad (10.13)$$

in a large volume of scintillator material placed near a nuclear reactor. Since fission fragments are neutron rich, the reactor emits mainly antineutrinos (cf. Eq. (10.4)). In another experiment, Davis looked for the emission of electrons in $\bar{\nu}_e$ interactions with nuclei but decided that

$$\bar{\nu}_e + Cl^{37} \overset{|}{\longrightarrow} A^{37} + e^-. \qquad (10.14)$$

This checks the conservation of electronic number N_e : e^- has $N_e = +1$, whereas $\bar{\nu}_e$ has $N_e = -1$.

(ii) a group at Brookhaven demonstrated directly, by a most ingenious experiment, that the neutrino emitted in K-electron capture in Eu^{152}

$$e^- + Eu^{152} \rightarrow Sm^{152} + \nu_e, \qquad (10.15)$$

does have negative helicity.

(iii) the observation of reactions induced by high-energy neutrinos. The neutrino beams used result mainly from the decay of intense, focussed, π^+ beams (with some K^+ particles included) at the 30 GeV proton accelerators. The striking discovery was made at Brookhaven that the reactions induced mostly involved μ^- production, typically

$$\nu_\mu + N \rightarrow P + \mu^-, \qquad (10.16a)$$
$$\nu_\mu + P \rightarrow \Delta^{++} + \mu^-. \qquad (10.16b)$$

This required at once that the neutrino v_e emitted with e^+ in beta-decay was different from the neutrino v_μ emitted with μ^+ in π^+ decay, a most remarkable conclusion.

This last conclusion had already been suspected from the evidence that

$$\mu^+ \nrightarrow e^+ + \gamma, \qquad (10.17)$$

despite very thorough search. If the μ meson were an excited state of an electron, in some sense, then it would be natural to expect some electromagnetic transition between them. Even if they were distinct particles, both with lepton numbers $N_l = 1$ and both coupled to the same neutrino, one would still expect this decay process to occur, since it is consistent with conservation of lepton number. Now we know that there are two distinct lepton numbers, N_μ and N_e and their separate conservation would be violated if this process (10.17) were to occur.

The cross-sections being measured in these high-energy neutrino experiments are very small. The typical cross-sections at these energies are 10^{-38} cm^2, to be compared with the pion-proton cross-section at high energies, which is about 3×10^{-26} cm^2; of course, this is just the ratio of weak decay rates relative to nuclear decay rates, the factor of about 10^{-13} which we mentioned at the beginning of this chapter. Nevertheless, these high-energy neutrino experiments provide a very natural and favourable way to look for the W boson, both because W production is more favoured in cross-section than the processes (10.16) provided the energy is sufficiently high relative to M_W, and because there is very little background in the detectors for these neutrino experiments. A typical production process would be

$$v_\mu + \text{Nucleus} \rightarrow \mu^- + W^+ + (\text{Nucleus})^*, \qquad (10.18)$$

where the connection with the nuclear particles is via the Coulomb interaction. The W^+ bosons would then decay through the following weak decay modes,

$$W^+ \rightarrow e^+ + v_e, \qquad (10.19a)$$
$$\rightarrow \mu^+ + v_\mu, \qquad (10.19b)$$
$$\rightarrow \text{pions}. \qquad (10.19c)$$

The most distinctive events would be those with decay mode (10.19a), for the reaction (10.18) would then lead to the simultaneous appearance of a μ^- and an e^+, which would be difficult to account for any other way. The decay mode (10.19b) would also be quite distinctive since the reaction (10.18) would lead to the emission of a (μ^+, μ^-) pair with wide angular separation. From the detailed analysis of a particularly thorough experiment at CERN, no clear-cut examples of these processes were found, and the number of "possible candidates" gave an upper limit on the W production rate, and therefore a lower limit on the W boson mass, $M_W c^2 \geqslant 2 \; GeV$.

With the interaction picture of *Figure 10-2*, the weak decay processes which do not involve leptons are considered to arise from the hadron currents interacting with themselves. Their relation with these hadronic currents is very much complicated by the effects of the strong interactions, and no really clear picture emerges. Decay processes such as

$$\Lambda \rightarrow P + \pi^-, \qquad (10.20a)$$
$$\Sigma^+ \rightarrow P + \pi^0. \qquad (10.20b)$$

are typical of the baryons and are known empirically to show strong P-violation effects; for example, the protons emitted from unpolarized Λ or from unpolarized Σ^+ are known to have a strong longitudinal polarization. These interactions can also give rise to meson decay processes which need either parity in the final state, for example

$$K_s^0 \rightarrow \pi^+ + \pi^-, \qquad (10.21a)$$
$$K^+ \rightarrow \pi^+ + \pi^+ + \pi^-. \qquad (10.21b)$$

It is interesting to mention that the interaction of these hadronic currents with themselves can give rise to $\Delta s = 0$ effects; for example, they will give a small parity-violating term in the nuclear force between two nucleons (although typically 10^{-6} smaller than the main terms) and there is good reason to believe that the effects of this tiny P-violation have recently been detected in some nuclei.

We turn now to our final topic about weak interactions, the properties of the neutral K-mesons. If we assume CP invariance, then we ask what effect the CP operation has on K^0 and $\bar{K}^0$:

$$(CP) \, K^0 \rightarrow - \, \bar{K}^0, \qquad (10.22)$$

since C changes particle to antiparticle and the parity P is $(-)$.

Then we have two states of special interest:

(a) $\qquad\qquad K_S^0 = (K^0 - \overline{K^0})$, for which $CP = (+)$ (10.23a)

(b) $\qquad\qquad K_L^0 = (K^0 + \overline{K^0})$, for which $CP = (-)$ (10.23b)

The final state $\pi^+\pi^-$ of Eq. (10.21a) has $CP = (+)$ since both C and P simply interchange the pions and their intrinsic parity is $(-)^2 = (+)$. The state K_S^0 can decay to 2π, whereas K_L^0 can only decay to more complicated states, such as 3π, which have much smaller energy releases. Hence K_S^0 has the short lifetime, and K_L^0 the long lifetime, of those listed under neutral K meson in Table I, and that is the reason for the subscripts used. But the expressions (10.23) show us that neither of the particles K_S^0 and K_L^0 has a definite strangeness; both of these particles have a part with $s = +1$ and a part with $s = -1$. This peculiar situation leads to many striking phenomena which have held a very great interest for elementary particle physicists. Among other things, their study has led to a good determination of the very small difference in mass between these particles,

$$m_L c^2 - m_S c^2 = (3 \cdot 6 \pm 0 \cdot 1)\, 10^{-6}\, eV. \qquad (10.24)$$

About five years ago, Cronin and Fitch at Princeton decided to test directly the statement made above, that the decay $K_L^0 \to \pi^+\pi^-$ could not occur. Essentially, their experiment was a direct and sensitive test of the CP-invariance which had been assumed to hold for all interactions. To general surprise, they found that this decay

$$K_L^0 \to \pi^+ + \pi^- \qquad (10.25)$$

did occur; in fact, about $0 \cdot 2\%$ of K_L^0 mesons decayed through this mode. The ratio which is most usually quoted in this connection is the ratio of the decay rates R,

$$\frac{R\ (K_L^0 \to \pi^+\pi^-)}{R\ (K_S^0 \to \pi^+\pi^-)} = (3 \cdot 6 \pm 0 \cdot 2) \times 10^{-6}, \qquad (10.26)$$

which illustrates well the smallness of this CP-violating effect. CP-violation has also been established for the neutral mode,

$$K_L^0 \to \pi^0 + \pi^0 \qquad (10.27)$$

but the values observed for this process for the ratio corresponding to (10.26) have varied widely, from zero to $13 \pm 3 \times 10^{-6}$ in the six experiments reported so far.

I do not think we have fully comprehended the implications of this result. Using this *CP*-violating process (10.25) it is now possible to specify an experiment which can determine for any observer whether the matter around him is made of particles or antiparticles. The characterization of an object as being particle or antiparticle is now *absolute,* and that carries the implication that we are now able to make an absolute distinction between a right-handed and a left-handed reference frame. The experiment is as follows. Take a beam of neutral K mesons which have well-defined strangeness at their point of production, for example, take a charged K beam and scatter it by some target to give neutral K-mesons by the charge-exchange process. Measure the time the neutral K-meson travels for each $\pi^+\pi^-$ decay event in this beam. This time, distribution consists of three terms. The first two terms are the exponential distributions due to K_S decay and to K_L decay separately, which fall off like $\exp(-t/\tau_S)$ and $\exp(-t/\tau_L)$ respectively; the third term is an interference term which arises from the fact that K_S and K_L are both coupled with the $\pi^+\pi^-$ state, and which has the time-dependence $\exp(-t/2\tau_S - t/2\tau_L)$. Look at the sign of this interference term for times t approaching zero. If it is positive there, then the neutral K meson produced was a K^0 meson (strangeness $s = +1$) and the charged K meson which gave rise to it was a K^+ meson; if negative, the neutral K meson was a $\overline{K^0}$ meson, and the charged K meson was K^-. This procedure can therefore be followed through by any observer for him to decide from the laws of physics whether the objects around him are made of matter or of anti-matter.

There are three possible interpretations of these experiments:

(i) *T*-invariance fails for some interactions, but in such a way that *CPT*-invariance does hold good. Lee has suggested that *T*-invariance may fail for the electromagnetic interactions, pointing out that the magnitude (10.26) of the *T*-violating effects is about equal to $(e^2/\pi\hbar c)^2$, which would be characteristic for an electromagnetic effect. Consequently, a number of experiments have been (and are still being) done to test *T*-invariance in photon processes (or processes where electromagnetic effects are crucial, even though no photon is emitted or absorbed). No firm conclusion has been reached yet about *T*-violations in other weak interactions, although a number of experiments involving various physical situations have failed to find any direct evidence for such effects, at least to the 5% accuracy level.

(ii) T-invariance may hold good, but CPT-invariance fails for some interactions. Here we need direct tests for CPT-invariance. One line of attack has been to measure accurately the ratio of lifetimes $\tau(K^-)/\tau(K^+)$ and $\tau(\pi^-)/\tau(\pi^+)$, since it is CPT-invariance which leads us to believe that these should be equal.

(iii) Both T-invariance and CPT-invariance may fail for some interactions. If both the $K_L^0 \rightarrow \pi\pi$ amplitudes were known completely, they would provide some direct tests, both for T- and for CPT-invariance. At present the data on $K_L^0 \rightarrow \pi^0\pi^0$ are too incomplete and uncertain for these tests to be significant and it is clear that there are going to be many more experiments on this decay process over the next few years.

There is one possibility of special interest which I want to mention. We know that there are no $\Delta s = \pm 2$ interactions of weak decay strength. We deduce this from the fact that the K_L^0-K_S^0 mass difference is of order G^2 (i.e., of order 10^{-14} of the meson masses) rather than of order G (which would give a mass difference about 10^7 times larger than that observed). Both K_L^0 and K_S^0 have parts with strangeness $+ 1$ and $- 1$, so that, if a $\Delta s = \pm 2$ interaction energy existed, it could connect the K^0 and $\overline{K}^0$ states and so generate a mass difference of the order of magnitude of its coupling amplitude G'. Wolfenstein pointed out that it would be quite possible for all the CP-violation effects to be confined to such a *hyperweak interaction* with $\Delta s = \pm 2$. To produce the observed CP-violating effects $K_L^0 \rightarrow \pi\pi$, this new interaction need have strength G' only of order 2×10^{-3} times G^2, which is the interaction strength appropriate to the K_S^0-K_L^0 mass difference. The effect of such an interaction would be completely negligible everywhere except in producing a CP-violating admixture to the K_S^0 and K_L^0 states; the weak interactions giving rise to the decay of these states could still themselves be completely CP-invariant. This possibility does require that the ratio between the $\pi^0\pi^0$ decay rates for K_L^0 and K_S^0 should have the same value (10.26) as for the $\pi^+\pi^-$ modes, and the present data still do not exclude this.

CHAPTER 11
(R. H. Dalitz)

Hunts for the Quark

The discovery of a free quark would represent a major turning point in "elementary particle physics" which would change its course abruptly, just as the discovery of the pion diverted the mainstream of attention away from the problem of nuclear forces in nuclei to the properties of the mesons whose interactions presumably gave rise to those nuclear forces, or as the discovery of strange particles pointed to a previously unsuspected dimension in "elementary particle" phenomena. A variety of opinions have been expressed about the likelihood that free quarks might exist and be found. Roughly speaking, these views may be summarized as follows:

(i) *quarks, a mathematical fiction.* Perhaps the quark variables, including spin, have only an abstract meaning, and the quark model itself is merely an artifice to simplify calculations, which could be carried out anyway through purely mathematical operations. This point of view was held rather generally at the time the "Eightfold Way" was first proposed; necessarily so, since no physical realisation had been proposed for the variables introduced with the octet model. This did not prevent calculations on the properties of the observed states from being made, and very successfully, using the methods of mathematical group theory. This view has become somewhat less palatable now that the introduction of "quark spin" and "internal orbital angular momentum L" has introduced a great deal of systematic order into the new spectroscopy. The very motivation for introducing these new elements into our supermultiplet schemes has largely stemmed from taking the quark model seriously in thinking about hadronic states.

451

(ii) *no free quarks*. Perhaps quarks and antiquarks can exist in the bound states which we observe as mesons, baryons and antibaryons, and an isolated, free quark is simply an impossibility. For example, crystalline solids behave exactly as if there were particles called "phonons" within them, which carry energy and momentum and which scatter from impurities and from each other; but the notion of a free "phonon", a phonon outside the crystal, is without meaning. This analogy is of interest, but its relevance for the elementary particle situation is not so clear. Another frequent suggestion is that quarks are bound by forces which are such that the quarks cannot be separated; this raises problems about how the quark compounds, the mesons and baryons, then manage to come together sufficiently to interact as we see them do.

(iii) *the naive view*, that the patterns of new spectroscopy point rather strongly to the interpretation that there exist some subunits within meson and baryon states, and that the regularities we have seen above have to do with the properties and interactions of these subunits. Then why are free quarks not seen in Nature? — perhaps because they are rather heavy (say 5-10 GeV), and require very high energy interactions for their production. And perhaps also their production cross-sections are very small — this is certainly the case for those theories which regard elementary particle collisions as giving rise to a hot blob of hadronic matter which then emits particles just as a hot liquid evaporates molecules, for evaporation is governed by a Maxwellian law which predicts an exponential fall-off in the probability of emission with increasing energy of the emitted particle. One particularly thorough series of calculations by Hagedorn at CERN suggests that the quark production rate predicted should fall by a factor 10^5 for each GeV increase in the mass of the quark.

The searches for quarks base themselves essentially on these last remarks, on the hypothesis that the quark is a heavy object, with mass measured in GeV. Not all of the quarks will be stable, of course. We expect the λ quark to be the heaviest (since particles with negative strangeness are generally heavier than their non-strange counterparts), so that it decays weakly to $p\pi^-$ or $n\pi^0$, or to $n\gamma$ if the mass difference is too small for pion emission.

Beta-decay may occur between the p and n quarks, depending on their mass relationship; one of them will be completely stable. Its antiparticle will also be stable, so we can expect at least two of these objects to be stable, one positively charged and the other negatively charged, with charges $\pm\ 2e/3$, or $\pm\ e/3$, or both.

There are three types of search which have been reported:

1. *Cosmic ray experiments.* Most of these experiments have looked specifically for particles of anomalously low ionisation compared with that for the majority of cosmic ray particles, with ionisation down by the factor $4/9$ for charge $\pm\ 2e/3$, or $1/9$ for charge $\pm\ e/3$. One experiment was designed to look for heavy particles (but sensitive only to $\pm\ e$) by requiring a time delay in their arrival at the detector relative to the shower from the interaction in which it was formed, since such a heavy particle (mass 5-10 *GeV*, say) would have a significantly lower velocity than the shower particles. The result of these experiments is usually stated as a quark flux, N_q quarks arriving per cm^2 of horizontal area, per unit solid angle centred on the vertical, per second. The best results available for either charge value are

$$N_q \leqslant 10^{-9}/cm^2 \text{ sr. sec.} \qquad (11.1)$$

A brief diversion on production thresholds is desirable here. In the laboratory frame, what we have generally is a particle of mass m and energy E (momentum given by Eq. (2.2)) incident on a stationary target of mass M. What really matters is the total energy E_c in the barycentric frame. If we want to create a quark pair, this is possible only if E_c exceeds $(2M_q + m' + M' + .\ .)c^2$, where m', M' are the masses of any other particles which are necessarily present in the final state. By considering the invariant $(E^2 - c^2P^2)$, as before, we conclude

$$\begin{aligned} E_c^2 &= (E + Mc^2)^2 - c^2P^2 \\ &= m^2c^4 + M^2c^4 + 2Mc^2E \end{aligned} \qquad (11.2)$$

We see that E_c rises much more slowly than the laboratory energy E, in fact only like $\sqrt{E}$ when E is large relative to Mc^2. For quark mass 10 *GeV,* and nucleon-nucleon collisions, then the threshold energy for quark pair production is $E_c = 22\ GeV$; from Eq. (11.2), we find that the corresponding laboratory energy required for the incident nucleon is about 240 *GeV*.

To return to the result (11.1) for N_q, there are two points to note about its interpretation. First, the flux to be expected depends on the assumptions made about the quark-pair production cross-section. If it is assumed that this cross-section has the constant value $\bar{\sigma}$ for all energies E above the threshold energy, then this flux (11.1) corresponds to the following limits.

$$M_q c^2 = 5 \; GeV \qquad\qquad 10 \; GeV$$
$$\bar{\sigma} \leqslant 2 \times 10^{-32} \; cm^2 \qquad 10^{-31} \; cm^2$$

These are quite weak limits, in comparison with the cross-section estimates made by Hagedorn (who would predict a cross-section of order $10^{-40} \; cm^2$ for $M_q c^2 = 5 \; GeV$). Second, it is important to realise that protons of very high energy are relatively rare in cosmic rays. If $M_q c^2$ were 10 GeV, the threshold proton energy would be about 240 GeV, as we said above; in all of these cosmic ray experiments, the total number of incident protons which had energy above the required threshold was only about 1000.

2. *geochemical search.* These assume that the quarks either come with the cosmic rays falling on the earth, or are produced by cosmic ray interactions in the earth's atmosphere. These quarks will be quite penetrating, because of their large mass, and we can expect them to come to rest in the earth's crust, or in the sea. We expect the negatively charged quark (or anti-quark) to be captured into an atom and to cascade down through the Bohr orbits, reaching ultimately some deep and stable atomic orbit. This atom now has, effectively, a nucleus with fractional charge. Since the electron charge is integral, we now have an atom with a fractional charge and this property can be used to select out the interesting atoms in a sample. Before discussing the experiments, let us estimate what would be a meaningful estimate for the density of quarks per nucleon, on the basis of the flux (11.1) and the assumption that all the quarks come to rest in the first 300 kgm/cm^2 of the Earth's crust. Assuming this has been going on for 5×10^9 years, the density estimated is 3×10^{-24} quarks/nucleon.

In an extensive series of experiments at the Argonne National Laboratory (Illinois) the procedure was to pass the material under test (air, sea water, or dust) past a plate electrically charged to attract and collect these peculiar atoms. Afterwards, the collectors were heated in order to evaporate off these special atoms to be re-collected more conveniently and analysed (for example) by a mass spectrometer. The greatest uncertainty lies in our lack of knowledge about the chemistry, and especially the surface chemistry, appropriate to these peculiar atoms — the assumption that they will be re-emitted by the heating applied is somewhat uncertain, for example. For sea water, the concentration of quarks was estimated to be less than 10^{-24} - 10^{-27} quarks/nucleon, the various limits being given for differing experimental conditions not described in detail in their report. For air, values of 10^{-30} to 10^{-33} quarks/nucleon were quoted, but it would not be surprising if the concentrations were especially low for air, since it would not be very effective in stopping the quarks.

Another group has developed an accurate technique for determining how much the charge on a diamagnetic speck differs from an integer. The speck is levitated in a magnetic valley. A known electric field is applied transversely and the displacement of the speck is determined. The charge on the speck is varied in integral steps by using a weak γ-ray source to liberate electrons from the speck. These integral steps serve to calibrate the equipment. If the speck has an integral charge, a point will be reached when the speck has zero charge and so suffers zero displacement by the electric field — if such a situation does not occur, then measurement of the smallest displacements gives the fractional value of the charge on the speck. So far all specks examined have had integral charge, but this means only that there are less than 5×10^{-19} quarks/nucleon in these specks of carbon.

The first problem with these searches is that one really needs to understand better the physico-chemical situation for quark-atoms, to know what are the most favourable locations or materials in which to carry them out.

3. *Accelerator experiments.* Here the problem is that the quark mass range which can be explored is strongly limited by the maximum proton energy. With 30 *GeV* protons at CERN and Brookhaven, the threshold is reached already for $M_q c^2 \approx 3 \ GeV$, unless one takes advantage of the Fermi motion of the nucleons within a complex nucleus in the target material; this mass limit

goes up to 5 *GeV* with the 70 *GeV* accelerator now in operation at Serpukhov (U.S.S.R.).

The most demanding experiment is one completed at CERN recently. Here a magnet system was set up to accept secondary particles emitted forward from a target irradiated by the proton beam. The magnet parameters were chosen to correspond to a momentum for a particle of charge $-e$ which was beyond the accelerator energy (this beam was known generally as the "super-momentum beam"); however, the magnet system would still accept particles of charge $-e/3$ with permissible momentum. This arrangement reduced the background enormously; then many detailed checks were also made to avoid possibilities for "fake events" to register. The quark pair production cross-section limits obtained in the CERN experiment were as follows:

Charge	$-2e/3$	$-e/3$	$+e/3$
σ_q	$\leqslant 10^{-36}\ cm^2$	$\leqslant 10^{-38}\ cm^2$	$\leqslant 10^{-35}\ cm^2$

For charge $(-e/3)$, this limit is below the value estimated by Hagedorn, after he had chosen a free parameter to give the observed cross-section for antiproton production, and his estimate should really be considered as rather pessimistic. This experiment makes it clear that if there are stable quarks with charge $-e/3$, their mass must certainly be greater than $2 \cdot 5$ *GeV*.

A preliminary experiment at Serpukhov has given a comparable cross-section limit for charge $-2e/3$; for quark mass as high as $4 \cdot 5$ *GeV*, however, an improvement by several orders of magnitude will be needed to bring the empirical limit down to the level of Hagedorn's estimate. The reason is as we gave earlier above; the production rate predicted falls exponentially, by a factor about 10^{-5} for every 1 *GeV* increase in M_q.

Physicists will continue to search for the quark, because of the enormous significance its discovery would have. However, unless quarks are found in the next round of experiments, the outlook is not bright. The energy of 300 *GeV*, typical for the proton accelerators now being designed, is quite sufficient for the production of quarks of mass 10 *GeV*, but the production cross-sections could well be so low that the quarks would never be detected.

CHAPTER 12
(R. H. Dalitz)

Survey of the Basic Forces in Nature

In this chapter, we shall survey briefly what we have learned up to ·this point and try to look for some overall pattern. In Table II we have listed all the categories of interactions we have had to distinguish in our discussion of "elementary particle" properties, together with the conservation laws and symmetry principles they have been found to obey. For each category of interaction, we have given an energy which is characteristic of its strength.

There is one important force listed there which we have not yet referred to, the force of *Gravity*. This is the dominant force when we look at the Universe as a whole, yet we do not mention it when we speak about elementary particles. The reason is that the gravitational energy is proportional to the product of the masses,

$$U(r_{12}) = \frac{-\,G\,M_1 M_2}{r_{12}}. \tag{12.1}$$

where this G is Newton's gravitational constant, with value $6\cdot67 \times 10^{-8}$ dyne cm^2/gm^2. The elementary particles have very small masses. For example, to take a favourable case, we may calculate the gravitational energy between all the nucleons of He^4, the result being about $1\cdot5 \times 10^{-35}$ eV.

Elementary particle physicists have given some attention to the graviton, the quantum of the gravitational field which will have spin $(2+)$, if Einstein's Theory of Gravitation is correct. The graviton field is treated in rather the same manner as the photon field. It is a gauge field coupled with mass-energy, and since the graviton field carries energy, the graviton field is coupled with itself, giving rise to non-linear equations of motion for this field. These equations of motion, derived from a completely different line of thought to that followed by Einstein, turn out to give exactly the equations of General Relativity; their interpretation is rather different since the basic tensor $g_{\mu\nu}$ which enters is interpreted simply as the potential of the spin $(2+)$ graviton field. The graviton is expected to have zero rest mass, and to travel with

457

TABLE II

Symmetry or Conservation Law	Name of Interaction					
	Super-strong	Strong	Electro-magnetic	Weak Decays	Hyper-weak (?)	Gravity
Unitary Symmetry	$SU(6)$	$SU(2)$	None	None	None	None
Space Reflection (parity P)	Yes	Yes	Yes	No	?	Yes
Particle - antiparticle conjugation C	Yes	Yes	Yes (?)	No	?	Yes (?)
Time Reversal T	Yes	Yes	Yes (?)	No (?)	No	Yes
Hypercharge Y	Yes	Yes	Yes	No	No (includes $\Delta Y = \pm 2$)	Yes (?)
Baryon No. B	Yes	Yes	Yes	Yes	Yes	Yes
Charge Q	Yes	Yes	Yes	Yes	Yes	Yes
Energy characteristic of interaction	$\sim 10^{10} eV$	$\sim 3 \times 10^{8} eV$	$\sim 5 \times 10^{6} eV$	$\sim 10^{-8} eV$†	$\sim 10^{-11} eV$	$\sim 10^{-34} eV$‡

TABLE II. The symmetries and conservation laws appropriate to each of the six categories of elementary particle interaction we have recognised. The inclusion of hyperweak is completely speculative. The notation ? means that the situation is unknown; (?) means that the corresponding item is not settled, but that experimental tests are going on to test this point. The entry † represents δm_K, the neutral K-meson mass difference, which may be a misleading measure of the weak interactions (see text). The entry ‡ gives the gravitational energy of the nucleons in He^4.

the velocity of light; it has not yet been detected and this is not surprising, since its coupling to the proton has strength $GM_P{}^2/\hbar c = 0\cdot6 \times 10^{-38}$, to be compared directly with the photon coupling strength $e^2/\hbar c = 1/137$.

In Table II, we note the enormously wide range of interaction strength spanned by the six categories of interactions we have distinguished. The hyperweak interaction is purely speculative at this point and could well have been excluded by the time these lectures are given. It is perhaps unreasonable to choose the K_S^0 - K_L^0 mass difference δm_K, which is proportional to G_W^2, as the measure of the weak interaction strength. If there does exist a W meson which has coupling strength $f^2/\hbar c$, a more reasonable energy to choose as characteristic of the weak interaction strength would be the mass differences generated by the interaction of particles with the W fields. These self-energies would be proportional to $f^2/\hbar c$, hence proportional to G_W; perhaps the geometric mean of δm_K and the strong interaction, i.e., about $1\ eV$, would be a more representative choice.

Next, we note that, looking aside from Gravity, the number of symmetry properties and of conservation laws satisfied by the class of interaction appears to increase in the same sequence as that for increasing strength of interaction. On the other hand, even though $SU(3)$ symmetry fails for electromagnetic and weak interactions, the $SU(3)$ symmetry still has definite implications for these interactions since they break the symmetry in a way which has some particularly simple form when expressed in terms of $SU(3)$ operations. Also, the weak interactions show a degree of organisation which is hard to express as a formal symmetry, but which may be due simply to the existence of an intermediate weak vector boson W coupled universally with the various weak currents; the situation depicted in *Figure 10-2* would lead directly to the lepton conservation laws $\Delta N_e = 0$ and $\Delta N_\mu = 0$, which we have not specified on Table II. The hadronic current which participates in this universality has a form which is peculiarly skew with respect to strangeness (due to the non-zero value for the Cabibbo angle θ given in expression (10.8)). Gravity does not really fit into this hierarchy pattern, but this may partly be because we know so little about gravitational forces.

CHAPTER 13
PART 1

(W. K. H. Panofsky)

The Future

A. Higher Energy Through New Technology

During Chapters 4 and 5 we described the evolution of accelerator design, resulting in an almost explosive increase in attainable energy. This growth has not been bought cheaply; the cost of modern proton accelerators is about one million dollars per GeV. There is no real end in sight at which the search for higher energy will be over. In the past each new step in energy has uncovered a new world of unexpected phenomena, and in all likelihood this will continue to be true. The high cost combined with the open future of the quest toward high particle energies have spurred much thought and experimentation on how to build higher energy accelerators for less. At this point there are two main directions in which this effort is going: (1) superconductivity, and (2) "collective" acceleration. Let me explain these approaches.

Certain metals become superconducting at very low temperature, i.e., they have no dc resistance at all below a certain transition temperature. It appears that application of this phenomenon might drastically reduce the cost and size of conventional magnet structures. The size of ordinary magnets is essentially controlled by the properties of the iron that is used. Iron saturates at magnetic fields of about 20,000 gauss, and at this field value a 300 GeV particle is bent through a radius of 500 metres. Therefore, a 300 GeV circular accelerator must have a circumference of at least two miles. Thus it is the magnetic properties of iron which force the size of present day accelerators to become quite large.

If an electromagnet is built without iron, i.e., with only ordinary coils carrying current, then the power loss in the coils for a large accelerator become prohibitively large. Hence an obvious answer appears to be an accelerator with superconducting windings. Small superconducting coils have been built to produce intense magnetic fields in the 100,000 gauss region as compared to the 20,000 gauss level of ordinary magnets as limited by the saturation of iron. Thus a reduction in radius up to a factor of five for a synchrotron, combined with reduction in power cost, appears feasible.

Unfortunately, the situation is not quite that simple. For one thing the special materials (usually alloys of niobium) which become superconducting at reasonable temperature, and which do not lose their superconducting properties at high magnetic fields, are quite expensive. Also, the saving in electric power is partially offset by the cost of the refrigeration needed to maintain the required low temperature against the unavoidable heat leaks. Finally, one must remember that the alternating gradient synchrotron as described in Chapter 4 uses a changing rather than a steady, magnetic field. It turns out that for changing currents super-conductivity is not complete: the changes in magnetic field produce some power losses which have to be removed by the refrigerator at low temperatures. It appeared for a while that this effect was so large that all the advantage of using superconductivity would be nullified. However, a ray of hope has recently appeared: When the conductors are made of very fine niobium-alloy wire which is continually transposed, then these power losses are very much reduced.

It is too early to predict when the first practical superconducting synchrotron will be built, or how much cost will be saved. It is likely, however, that if machines of energy much higher than 300 GeV (the energy of the planned European accelerator) are to be built, they are likely to be superconducting.

A similar possibility exists for the linear electron accelerator. You will recall from Chapter 4 that all present-day linear accelerators are pulsed machines, because otherwise the high power dissipation in the walls could neither be supplied nor cooled Superconductivity changes all that. In principle it should be possible to build a linear accelerator with superconducting walls which operates continuously and still have the refrigerator consume only about the same amount of electric power as the radio-frequency power tubes now do in present-day linear accelerators (such as the Stanford two-mile machine, described in Chapter 4). Moreover, it should be possible to obtain perhaps five times as much energy in a given length. However, here again there is a gap between present achievement and future expectation. At radiofrequencies a superconductor does not have zero electrical resistance; theoretically the power losses only go to zero as the temperature goes to absolute zero ($-273 \cdot 1^{0}C$ or $0^{0}K$). Even this is not true in practice: Due to the presence of imperfections, the losses do not decrease further with decreasing temperatures below a certain value. Tests on small test cavities have shown that, with care in processing and handling the superconducting material (pure niobium is most promising in this case), the deviation from ideal behaviour can be made small enough that the expectation of building a linear accelerator with continuous beam at very high energies looks most promising.

A further existing possibility is the ERA or "*e*lectron *r*ing *a*ccelerator". This is the outgrowth of an old idea, originally due to the late Soviet physicist, Vladimir Veksler (the co-inventor of the phase stability principle discussed in Chapter 4). The root of the idea is the following. If a cloud of negative electrons is accelerated, what happens to a positive proton that is carried along by the electron bunch? If enough electrons are present then the attraction between the electrons and the proton would "bind" the proton to the electron bunch. If, therefore, the proton and the electron move with the same speed, the energy of the proton is 1836 times that of each electron since the proton has a mass 1836 times that of the electron. Therefore one might guess that if a lonely proton got stuck in the electron bunches of Stanford's 20 *GeV* electron linear accelerator, it might have an energy of 36,720 *GeV*!

Now what is wrong with this idea? (And it is wrong!) The trouble is that the force required to hold the proton in the electron bunches while they are being accelerated to this fantastic energy is very large, and it could be achieved only if there were very many electrons very close together. It turns out that a cloud of electrons sufficient to hold the proton under these conditions would blow itself apart due to the electric repulsion among the negative electrons. However, there appears to be a way to beat this game: This is to form the electron bunch not as a diffuse cloud but as a ring rotating at high speed. Such an electron ring will hold itself together since now the electrons are not only a bunch of *charges* repelling one another electrically, but are also a collection of current loops which attract one another magnetically. I am sure you know that two wires carrying a current in the same direction will attract one another. It can be calculated that at high speeds the electric repulsion and magnetic attraction cancel almost exactly; therefore one can pack a lot more electrons close together into such a ring than into an ordinary bunch.

This principle actually works: experiments at Berkeley, U.S.A., and in Dubna, U.S.S.R., have led to the formation of such an electron ring in which protons have been captured. Many questions remain: Can such a ring be accelerated without losing its precious cargo of protons? And, supposing acceleration works, how much does it cost? One should note that the motivation for pursuing new accelerator ideas is economic: Given enough money one can technically build accelerators of the type described earlier in Chapters 4 and 5 to any arbitrary energy; therefore the "proof of the pudding" of the new accelerator ideas is not so much whether they work, but whether they get to higher energy at lower cost. Some answers to this question will be known in a few years.

B. Higher Useful Energy Through Colliding Beams

There is another road to higher energies using a more subtle approach. As we discussed in Chapter 4, not all the energy of a bombarding particle hitting a target nucleus is "available" for energy release in the collision; some energy is simply transformed into energy of "motion of the centre-of-mass", that is the energy in moving the entire colliding mass forward. Clearly this energy is not useful; only the energy seen by an observer moving along with the common centre-of-mass of the colliding particle can be released for producing interactions. *Figure 1-8* of Chapter 1 plots the available "centre-of-mass" energy against the laboratory energy of the bombarding particle. For example, the 200 *GeV* accelerator will produce only 19·5 *GeV* of available energy. In

Chapter 1 we introduced the approximate formula, Eq. 1.7, valid at very high energies, which we write here as

$$E_{<\text{centre-of-mass}>} = \sqrt{2M_0c^2E_{<\text{laboratory}>}} \qquad (13.1)$$

where M_0 is the rest-mass of the target particle. If we would let two particles each of energy $E_{<\text{laboratory}>}$ make a head-on collision, then the "centre-of-mass" is at rest and we would simply have

$$E_{<\text{centre-of-mass}>} = 2E_{<\text{laboratory}>}. \qquad (13.2)$$

If, therefore, two 200 GeV beams were circulating in an accelerator in opposite directions, and were then to make a collision, the available centre-of-mass energy would be 400 GeV. To obtain the same collision energy by striking a target proton at rest, the required beam energy would be 85,000 GeV (!). A "storage ring" in which 200 GeV beams circulate in opposite directions is certainly cheaper than an 85,000 GeV accelerator.

From the energy point of view a colliding beam device is thus clearly a great advantage. From the intensity standpoint things are less favourable. A material target, for example a bottle of liquid hydrogen, contains about 4×10^{22} protons/cm^3. An intense high energy beam of protons, say of 1 ampere, which has transverse dimensions of $0 \cdot 1$ cm by $0 \cdot 1$ cm has a density of 2×10^{10} protons/cm^3 only, or a factor of 2×10^{12} less! In fact the density of particles in a beam is often lower than the density of molecules in the vacuum chamber of the accelerator. Therefore the attainable data rate in a colliding beam apparatus is much less. Such a device could thus never replace the conventional accelerator in which a stationary target is hit, but because of the tremendous energy advantage it offers a "window into the future" to observe and measure at least the more copious processes which would occur at extremely high energies.

The promise of this technique is so large that the European laboratory (CERN) at Geneva is building a huge "intersecting storage ring" installation in which two beams of 25 GeV protons circulate in opposite directions. *Figure 13-1* shows a diagram of the installation. If this technique proves successful at CERN, then it is likely that a good case can be made to add storage ring installations to the proton synchrotrons operating at even higher

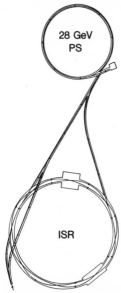

Figure 13-1. Diagram of the intersecting storage rings (ISR) being built at CERN, in Geneva, Switzerland.

energies, such as the planned 200 *GeV* accelerator near Chicago, Illinois, U.S.A.

Successful storage rings have already been built for electrons and positrons. Here the scientific incentive is somewhat different and is not primarily related to achieving very high centre-of-mass energies. The reason is the following: Since, as discussed in Chapter 4, high energy electrons lose their energy rapidly when circulating in a magnetic field, storage rings for electrons become impractical at very high energies, say above 4 *GeV*; in that case the centre-of-mass energy is 8 *GeV*. If, on the other hand, the 20 *GeV* beam of the existing Stanford accelerator strikes a stationary target containing hydrogen, the centre-of-mass energy is 6·1 *GeV* which is not too different from the practical maximum attainable with an electron storage ring. The singular potential of a storage ring for electrons and positrons lies elsewhere: It is a source of the "purest" energy known. Let me explain: Let me assume that a storage ring has been built into which a linear accelerator has injected electrons (e^-) and positrons (e^+) circulating in opposite

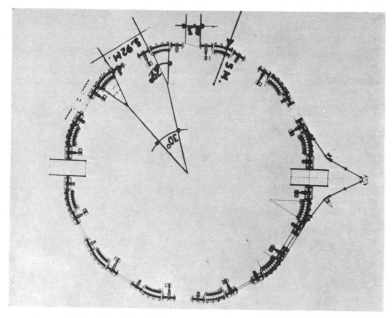

Figure 13-2. The 1.5 GeV electron-positron colliding-beam storage ring at Frascati, Italy.

directions. When the particles collide they can annihilate one another releasing energy in various forms. For instance we can visualise reactions such as:

$$e^- + e^+ \rightarrow \text{energy} \rightarrow \pi^+ + \pi^-$$
$$\rightarrow \mu^+ + \mu^-$$
$$\rightarrow K^+ + K^- \qquad (13.3)$$
$$\rightarrow p + p^-$$

Study of these reactions provides a much simpler test of theory than conventional experiments where a proton at rest is struck by an energetic particle. This is because the target proton cannot be destroyed (the number of nucleons cannot be changed in a reaction), and so the initially present nucleon will still be around in the final state and will complicate the analysis even if some of the other reaction products are the ones being studied. For example, the reaction possible with a conventional accelerator

$$\gamma + p \rightarrow \pi^+ + \pi^- + p \qquad (13.4)$$

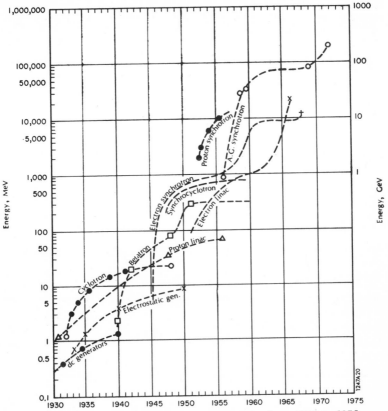

Figure 13-3. Energies achieved by accelerators from 1930 to 1975.

is a much more complex tool for studying the force between the π^+ and π^- than the reaction possible in storage rings:

$$e^+ + e^- \rightarrow \pi^+ + \pi^- \qquad (13.5)$$

Therefore electron-positron storage rings provide a new form of experimental technique in which the forces between pairs of unstable, short-lived particles can be studied without interference. The electron-positron storage rings constructed in Europe have already demonstrated their extraordinary usefulness in this respect.

Technically the construction of storage rings for electrons and positrons pushes technology to its limits in many respects: The particles must be injected into the rings efficiently without much loss; the energy lost by radiation must continually be replenished through reacceleration by radiofrequency power; the vacuum in the ring must be extraordinarily high (about 10^{-13} of atmospheric pressure) in order to prevent excessive beam loss. In such rings the electrons and positrons in the beams circulate for about one hour before the ring has to be refilled; in this time each electron has travelled about 10^{14} *cm*, or nearly one billion miles! This is several thousand times the distance from the earth to the moon and back.

Figure 13-2 shows a photograph of the electron-positron storage ring at Frascati, Italy; this machine is now starting operation at a beam energy of $1 \cdot 5$ *GeV*, the highest energy reached to date in this kind of device.

The quest for better and more powerful tools to provide high energy particles for studying the innermost structure of matter is thus far from being at an end. Thus far new techniques have continuously appeared on the horizon when the old methods appeared to have reached their limit. *Figure 13-3* shows how the attainable energy has climbed during the last decades; the curve shows little indication of slowing down in the future.

CHAPTER 13
PART 2

(R. H. Dalitz)

Open Questions

I hope that I have impressed you in these lectures with how little we understand about this field of physics, in comparison with the rapidly growing body of experimental information. At every turn there have been surprises. With little understanding, we are now facing further uprooting of notions about properties of space-time which we have thought recently to be both eminently reasonable and secure. So it is certainly a difficult proposition to try to anticipate what questions will seem important within the next five years, say. I shall content myself with listing a series of questions which it seems reasonable to ask today, together with some commentary:

Where is Asymptotia? How do high-energy reaction cross-sections behave as the energy increases indefinitely? Shall we reach an asymptotic region where all the interaction phenomena have settled down to a smooth dependence on the energy? Do meson-baryon and baryon-baryon total cross-sections approach constant values ultimately — or do they all approach the limit zero, but at different rates? Total cross-sections appear to vary quite slowly with energy above 20 GeV — have we already set foot in Asymptotia? Theoretical opinion has swung back and forth between these possibilities. These questions provide one of the major reasons for the construction of very high energy accelerators. For the proton-proton interaction at least, these questions will be answered in greatest detail by the Intersecting Storage Rings (ISR) at CERN, where 20 GeV colliding proton beams will reach conditions equivalent to 800 GeV protons incident on target protons.

Are there Heavy Particles?   Are there sub-units within the hadrons?   The new spectroscopy (and *Figure 2-1*) makes the case for some kind of hadronic substructure look rather strong.   Recent data on inelastic electron scattering from Professor Panofsky's laboratory (SLAC) at Stanford suggest that there may be some point-like objects within the proton.   In the section on quark hunts, we have discussed the difficulties facing a clear-cut answer to the question whether *heavy* sub-units do exist, and these difficulties persist whatever the nature of these sub-units.

Is there a carrier boson W for the Weak Interactions?   High-energy neutrino studies offer a very direct approach to this question and the availability of these high-energy v and $\bar{v}$ beams is another major factor which justifies the construction of high-energy accelerators such as the 200-400 *GeV* machines now in preparation.   Many hadronic interaction processes have also been proposed for $W^{\pm}$ searches, such as

$$P + P \rightarrow W^{+} + P + N, \tag{13.6}$$

$$\pi^{+} + P \rightarrow P + W^{+}, \tag{13.7}$$

where the production cross-sections may be as low as 10^{-7} times the values usual for strong-interaction processes (provided the W^{+} mass is low enough to allow W^{+} production at the energies available), with various procedures proposed to pick out the W^{+} events selectively.   An experiment of type (13.6) has been proposed for the ISR when it first comes into operation, and such experiments have already been considered for the 30 *GeV* accelerators.

New Particles to be found?   All the indications are already that there are very many particles still to be discovered for higher mass values and higher spin values.   But suppose we consider mesons with a definite set of values J, P, Y, Q — will there be an (essentially) infinite set of these particles, say for example of charged, non-strange, vector mesons?   These would correspond to an infinite multiplicity of Regge trajectories on *Figure 9-13*.   Most theories suggest this should be the case — in the quark model, they represent a sequence of radial excitations, in some other model they will have some other name and/or origin.   The electron-positron storage rings might have a special role to play here, in that e^+ - e^- annihilation reactions favour vector meson states and have already given us our most detailed knowledge of the ρ^0, ω and ϕ mesons we already know.

New Quantum Numbers?   We see no hint of this at present. Nevertheless, model calculations with theories having further additive quantum numbers have shown that these can play their role in quite a complicated way, not easy to unravel. The only point that is clear is that additional quantum numbers generally lead to much larger supermultiplets. At present, $SU(6)$ symmetry appears sufficiently complicated to account for the richness of particle states established so far. But it is a chastening thought that, without the cosmic radiation and if we were just reaching the sophistication of 750 MeV synchrocyclotrons, we would have had no hint yet of the strangeness quantum number.

T and/or CPT failure?   Does T-invariance hold for electromagnetic interactions? If so, then where does it break down, in which interactions? Or is it CPT-invariance which is being violated? If the latter, then how can one construct a theory for which this violation is the case? Tests for T and CPT exist from the $K^0_{L,S} \rightarrow 2\pi$ phenomena and we can expect definite statements from them within several years. Tests for C in electro-dynamics from the $\eta \rightarrow \pi^+\pi^-\pi^0$ process should also be available in the same time (present experiments show a 3 standard-deviation effect for C-violation).

Validity of Photon-Lepton Theory?   The system of electrons, photons and external fields is the only successful field theory we have today. This is partly because of the smallness of the expansion parameter $\alpha = e^2/\hbar c = 1/137$. It is true that the calculations lead to some infinite quantities, but we have learned how to extract the finite parts from these calculations, using the concepts of mass and charge renormalisation to give a unique definition and a physical interpretation to the infinite parts. Agreement between the theoretical calculations and the experimental data for the low-energy phenomena (e.g., lepton magnetic moments) is excellent up to the third order corrections in α, but these phenomena do not depend sensitively on the region of high momentum transfers. The electrodynamic theory has been tested very directly by observations on electron-electron scattering using colliding beams of 500 MeV; it has also been tested by accurate measurements of wide-angle e^+-e^- and μ^+-μ^- pair production by photons. The present conclusion is that the theory is certainly good up to momentum transfers of order 4 GeV/c (alternatively, down to distances of 5×10^{-15} cm) and could well be good for arbitrarily high momentum transfer. These tests will clearly be continued, as long as more searching tests are still possible.

More Leptons? Why do we have two distinct lepton families? The μ meson and the electron have exactly the same interactions, apart from the substitution of m_μ for m_e, wherever tested; the most searching test is the comparison between μ^-P and e^-P elastic scattering, which are considered to agree within errors up to about $0\cdot9$ GeV/c momentum transfer. The μ meson and its neutrino mate ν_μ appear superfluous — we do not understand the need for them, nor do we understand the large ratio $m_\mu/m_e \approx 207$. Are there excited leptons, e' and μ'? If so, they could give much-needed clues to the internal structure of the leptons and to their relationship. These excited leptons would probably undergo rapid γ-decay, $e' \rightarrow e + \gamma$, $\mu' \rightarrow \mu + \gamma$. Perhaps such excited electrons could be formed in electron-electron scattering, signalled by a high-energy γ-ray from their decay, or in high energy $\bar{e}$-e' pair production?

This list is far from exhaustive, and I am aware that I have not sufficiently stressed the importance of increasing the data shown on *Figures 9-4* and *9-8* about baryonic and mesonic states and of extending the data to higher mass regions.

The accelerators we have today are probably capable of extending our knowledge of hadron spectroscopy as far as we theoreticians are likely to need it, although beams of higher intensity would be a great advantage in many aspects of this work; clearly these accelerators will have a busy decade ahead of them. Particular questions are likely to receive much illumination from rather special accelerators such as the $1\cdot5$ GeV electron-positron colliding beams machine now coming into operation at Frascati (Italy), and the 3 GeV electron-positron colliding beams becoming available at the Cambridge Electron Accelerator (U.S.A.) a few years later, with perhaps even proton-antiproton storage rings by the end of the 1970's, as envisaged by Budker at Novosibirsk.

We shall also look forward to new kinds of data, and to work at higher energies, as the accelerators at present under construction come into operation during this next decade. The 20 *GeV* proton intersecting storage rings at the European Organisation for Nuclear Research (CERN) are expected to be operational within about two years and may be expected to teach us about the details of the proton-proton interactions up to energies equivalent to 900 *MeV* laboratory protons incident on target protons, a regime only tentatively explored through cosmic ray studies so far, and perhaps representing Asymptotia for these interactions, as we see things now. These experiments may well prove to be our best bet in the search for quarks, or for the *W*-meson. We also hope to see the Stanford Linear Accelerator (U.S.A.) in use for beam energy as high as 40 *GeV,* to extend our knowledge of the internal structure of nucleons both with greater spatial resolution and with greater precision. Later, there will come the 200-300 *GeV* proton alternating-gradient synchrotrons now being constructed both at Batavia (U.S.A.) and in Europe. These will be able to teach us about Asymptotia for other hadron-hadron interactions and reaction processes; perhaps even more important, the neutrino and anti-neutrino beams obtainable from the secondary beams from these accelerators will have energies almost an order of magnitude higher than those available at present, and we may look forward to learning a great deal more in detail about the weak interaction currents, and perhaps about the existence of the *W*-meson as well. Finally we should also mention that these last accelerators will also give us secondary meson beams of much higher intensity than those available today in the intermediate energy region so important for hadronic spectroscopy. We look forward to a much greater understanding of the problems that the "elementary particle states" pose for us during the next decade.